Evolving Insight

Evolving Insight

RICHARD W. BYRNE

OXFORD
UNIVERSITY PRESS

OXFORD
UNIVERSITY PRESS

Great Clarendon Street, Oxford, OX2 6DP,
United Kingdom

Oxford University Press is a department of the University of Oxford.
It furthers the University's objective of excellence in research, scholarship,
and education by publishing worldwide. Oxford is a registered trade mark of
Oxford University Press in the UK and in certain other countries

© Oxford University Press 2016

The moral rights of the author have been asserted

First Edition published in 2016

Impression: 2

All rights reserved. No part of this publication may be reproduced, stored in
a retrieval system, or transmitted, in any form or by any means, without the
prior permission in writing of Oxford University Press, or as expressly permitted
by law, by licence or under terms agreed with the appropriate reprographics
rights organization. Enquiries concerning reproduction outside the scope of the
above should be sent to the Rights Department, Oxford University Press, at the
address above

You must not circulate this work in any other form
and you must impose this same condition on any acquirer

Published in the United States of America by Oxford University Press
198 Madison Avenue, New York, NY 10016, United States of America

British Library Cataloguing in Publication Data

Data available

Library of Congress Control Number: 2015947529

ISBN 978-0-19-875707-8

Printed and bound by
CPI Group (UK) Ltd, Croydon, CR0 4YY

Oxford University Press makes no representation, express or implied, that the
drug dosages in this book are correct. Readers must therefore always check
the product information and clinical procedures with the most up-to-date
published product information and data sheets provided by the manufacturers
and the most recent codes of conduct and safety regulations. The authors and
the publishers do not accept responsibility or legal liability for any errors in the
text or for the misuse or misapplication of material in this work. Except where
otherwise stated, drug dosages and recommendations are for the non-pregnant
adult who is not breast-feeding

Acknowledgments

In writing this book, I have continually been reminded of my debt to the excellent PhD students, research assistants, and postdoctoral fellows with whom I've had the pleasure of working for many years. The ideas developed here were all worked out while working jointly with them, and—though their names occur often in the references—probably owe more to my collaborators than the raw text would seem to suggest! Their topics have spanned all the groups of animals that feature prominently here: prosimians (April Ruiz), monkeys (Peter Henzi, in collaboration with my colleague Andy Whiten; Rob Barton, Kate Harrison, Heather McKiggan, Mindy Babitz, Patti Teixidor, Rogério da Cunha, Alexandra Valero, Karline Janmaat, Rahel Noser, and Cris Cäsar), and great apes (gorillas: Jen Byrne, Joanne Tanner, and Emilie Genty; orangutans: Erica Cartmill; chimpanzees: Emma Stokes, Nadia Corp, Lucy Bates, Cat Hobaiter, Katie Hall, Hélène Cochet, and Brittany Fallon; and bonobos: Lisa Orr and Kirsty Graham); not to mention elephants (Lucy Bates, again, and Anna Smet) and pigs (Suzanne Held, in collaboration with Mike Mendl at the University of Bristol). Without you lot, I might not have had anything to write about: thank you! Special thanks are due to those who heroically read and gave me comments on the book itself: Cat Hobaiter and Lucy Bates, both of whom have been such fine colleagues for so long.

Contents

1 Introduction: What is insight, and why care about its evolution? *1*

2 Why are animals cognitive? The need for cognitive explanation in animal behavior *8*

3 In the beginning was the vocalization: Vocal communication in monkeys and apes *14*

4 Gestural communication in great apes *22*

5 Understanding others: Reacting to what others see and know *40*

6 Social complexity and the brain *52*

7 Learning from others: Cultural intelligence? *70*

8 Theory of mind: Understanding what others think about the world *84*

9 Pivot point: From social to technical abilities *108*

10 Knowledge about the physical world *117*

11 Learning new complex skills: Behavior parsing and the origin of insight *134*

12 A road map to insight: How it is we can think about why things happen *155*

References *169*
Author index *191*
Subject index *195*

Chapter 1

Introduction

What is insight, and why care about its evolution?

Insight is not a very popular word in psychology or biology, but that is actually partly the reason why I chose it for the title. Popular terms—like "intelligence," "planning," "complexity," or "cognitive"—have a habit of sprawling out to include everyone's favorite interpretation, and end up with such vague meanings that each new writer has to redefine them for use. Insight remains in everyday usage: as a down-to-earth, lay term for a deep, shrewd, or discerning kind of understanding. Insight is a good thing to have, so it's important to find out how it evolved, and that's what this book is about. (Strictly speaking, of course, insight doesn't evolve, it dawns. But that linguistic awkwardness is no better with the word "understanding," and with insight it will be particularly clear to readers—I hope—that what I mean is the evolution of the capacity for having insights.)

If you look past the everyday sense of insight, you soon meet Köhler's chimpanzee, Sultan. Sultan's behavior is the stuff of Comparative Psychology 101. During the Great War, when Wolfgang Köhler was interned on Tenerife and spent his time studying the captive colony of chimpanzees there, Sultan was his star subject. Köhler had presented chimpanzees with a range of tasks, including the use of a rake to get foods that were out of reach even to the long arm of a chimpanzee. Sultan was good at raking, but he was stumped when Köhler gave him several short sticks, none of them long enough to reach the food. After a while, Sultan gave up and simply sat next to some of the sticks. Then insight dawned. While fiddling with the sticks for no apparent purpose, Sultan happened to push two sticks together—and they held, making a longer stick. Sultan suddenly became animated, took the sticks across to the out-of-reach food, and used his new, combined tool to reach it, with immediate success! That is insight, as described in many psychology texts: the "Ah ha!" moment. Trouble is, it's not quite clear what took place for Sultan to give him the insight. Subsequent replications with chimpanzees reared in relatively deprived circumstances have shown that the revelation only happens to individuals that have had the chance, as children, to play with sticks and other objects. Long experience, as well as an "Ah ha!" moment, seem to be necessary. And what actually happens in that "Ah ha!" moment, anyway? It smacks of unconscious thought, and "explaining" an observed phenomenon with something even more mysterious is not satisfying.

In this book, I will treat insight as the ability to inspect and manipulate a mental representation of some part of the world, when away from any opportunity to see or in any other way to perceive it directly. "Mental representation" is a useful technical term in psychology: a mental representation of something in the world is the brain structure that enables us to think about it. For example, you know your way around your home very well: if asked, you can explain how to get from the lounge to the kitchen, what is directly below the spare bedroom, and what you can see from the dining-room window. That means, in psychological terms, you have a mental representation of the house, which you can not only use to direct your actions, but also "manipulate and inspect" mentally, in order to answer those questions when you are asked, even if you are nowhere near the house at the time. This representation need not—in fact, is most unlikely to—preserve every detail of the real world. We do not have photographic memories, and our knowledge even of our own homes is not equivalent to a set of architect's drawings; nor would it be useful to have such accuracy and completeness, since we'd be bogged down with trivia. What is important for a mental representation is that it encodes useful information, and can be manipulated in the mind—the process we call "thinking" in everyday talk. Insight is therefore about thinking.

To spell this out for Köhler's example: Sultan was able to think about his problem, even when not engaging with the task, because he had a mental representation of its structure in his mind, crucially including the need for a longer stick. In everyday language, when he discovered that two short sticks could be combined into one longer one, which he might never have done without an enriched childhood of play with similar objects, he "saw" its significance: the longer stick (in his hand) was relevant to the unsolved food problem. That is to say, more formally, that by combining the mental representation of the unsolved problem which he'd given up, with the evidence in his hands of how sticks can be combined to make a single longer one, he computed a solution to the problem—in his mind. Then, of course, he quickly returned to the site of the problem to enact the new idea, with success. Now suppose Sultan had solved the problem the other way around: suppose he'd sat and thought while remaining at the food site, and remembered his childhood play with sticks that had sometimes allowed him to make a longer one. That is also insight; in that case, the mental representation critical for success would be an insightful representation of his early experience and its results, including the principle of stick-extension that if recalled could be applied to the current problem. But had he remained at the food site, fiddling with the sticks and constantly re-trying the objects at hand, no mental representation would have been necessary to explain sudden success: trial and error might just have given him a useable tool, with no insight needed.

Taking this interpretation immediately offends almost everybody in my discipline of psychology.

Comparative psychologists have spent 100 years constructing learning theories that do away with the need for anything like using mental representations to explain behavior. Some, most famously the behaviorist B F Skinner, the towering figure in mid-twentieth-century psychology in the USA, even-handedly applied this approach also to human behavior—to the anger and disgust of philosophers and linguists. The furor eventually resulted in abandonment of the Skinnerian project, already undermined by abject failures to make computer-based models of language that worked using only learning theory principles. (Whether this outcome was entirely fair, given that no other computer models of language worked either, whatever their theoretical bases, is another story.) From around 1960 onward, human experimental psychology became cognitive psychology, a subject founded on mental representation. Cognitive psychologists study humans and they don't much bother with animals: the assumption is that the use of mental representations is unavailable to animals, most likely because mental representations are causally related to possession of language, which animal species lack. Meanwhile, comparative psychologists continue to apply animal learning theory to the study of animals, and they continue to avoid any talk of mental representation like the plague.

The suggestion that the capacity for insight evolved earlier than language—and so it may be shared with other living species—would be regarded as a pretty dangerous one by most cognitive and comparative psychologists. It is disturbing to cognitive psychologists, because the idea that insight has an ancient evolutionary origin would imply that human mental representations—and thus thinking—also had an evolutionary history in non-linguistic forbears. And if things didn't all start with language, maybe the roots of language predate the human lineage, too? It is disturbing to comparative psychologists, because it suggests that non-human animals may be able to think after all. These ideas may be shocking to psychologists, but what about to other people? Would non-psychologists be surprised at the idea of insight and thought in non-humans? As far as I can tell, it simply never occurs to most people—including most professional zoologists—that animals have anything qualitatively different inside their heads than humans do. If people compute with mental representations, then, presumably all animals must also do. So there is a risk that I may offend them too, by arguing in this book that most animals *lack* insight into how things work, including each other. Insight is something quite special. It may have evolved only in a few lineages, or perhaps even just our own.

To show insight, then, is more than merely responding appropriately or learning new actions. Insightful understanding implies possessing a mental representation or model of what is out there. The mental model can be used to work things out, by computing in the head: thinking, to use the everyday term. As a result, behavior is based on expectations about the future, rather than purely a product of what happened in the past.

Let us assume that most animals do not understand anything in this insightful way: which ones might be the exceptions? Our closest relatives, the primates, are an obvious possibility. After checking them, we might look at other species that strike observers as flexible and innovative in their actions: for example, rats and crows. Then there are animals that have surprisingly large brains, such as dolphins and elephants. Surely larger brains are for thinking? All these animals will feature in this book. I will not have much to say about lizards, ducks, and shrews—perhaps because they do not understand the world in an insightful way. Or perhaps they do, and simply have not attracted the serious attention of scientists yet: a challenge for the future.

An assumption, that large-brained animals are probably smart, has been slipped in and might have passed without comment: but is it safe? Parrots strike most people as rather smart, yet their brains are the size of walnuts, much smaller than those of many apparently "dimmer" mammals. Yet for a bird of parrot size, their brains are unusually large. Often, brain size is expressed in relation to body size. It is almost obvious that smaller animals are going to have smaller brains, on the whole. But should we expect their brains to be scaled in exact proportion to their bodies? In order to approach a useful understanding of brain size, I shall have to touch on the subject of allometry—the geometric scaling of the sizes of parts of living things as their overall size changes. I shall also need to bring in some computational theory—how exactly does a computational process work and what makes it more powerful—because it is the brain as on-board computer that matters for the ability to show insight.

To a cognitive theorist, the best way to describe intelligent behavior is in terms of what mental representations allow it to be done. In this framework, a lack of insight corresponds to an inability, when we are not actually performing an activity, to interrogate and manipulate the mental representations which record our perceptions and control our motor capacities when we're busy doing that activity. In everyday terms, a lack of insight corresponds to an inability to think, to go beyond simply seeing and doing, in order to work out other consequences. But is this framework really necessary, to describe what controls the behavior of animals that lack insight? Might an account that directly relates seeing to doing be better? It would certainly seem simpler.

The traditional view from psychology is that what animals do when they adjust their behavior to the world is "just" learning. The animal has no mental model of the world, but only a set of links between eliciting circumstances (how the world looks, smells, or sounds) and actions in its repertoire that have worked before. The links may be provided by genes. The first scientists to study animal behavior under natural conditions, like Konrad Lorenz, developed a theory that particular triggers may act to "release" innate patterns of behavior. A famous example of this releaser + fixed-action-pattern idea is the way that the red color of a robin's breast triggers an attack by a male robin. Alternatively, the links may be built up associatively, simply by occurring together in time and space ("classical conditioning," as Pavlov's dogs came to salivate at the sound of a bell that was always sounded when they ate), or as a result of past success in trial-and-error exploration ("instrumental conditioning," as the laboratory rat comes to press the bar in a Skinner box which usually leads to delivery of a sugar pill). Genes and learning may work together to better effect. For instance, rats are able to associate nausea with some novel food they ate, even after a long time lag. This phenomenon has been called "constraints on learning," but from the rat's point of view it is very valuable: a better description would be to say that genes channel learning toward biologically important relationships, like that between nausea and recent unusual foods. And the social context has the potential to aid learning powerfully. Many species have innate tendencies which focus their attention on the right places to explore and the right actions to try first. These mechanisms were all described in my previous book, *The Thinking Ape* (1995a), and will only be briefly treated here. If all these simple principles are invoked, an impressively wide range of animal behavior can be understood as "just learning." Why do we need more?

Without insight, reactions are guided only by what can be perceived at the time: that is, by surface features rather than understanding. Whether learned or genetically encoded, behavior driven entirely by linking perception and possible actions has the following features:

1 Causal, eliciting circumstances are entirely perceptual, e.g., they cannot include a structural account of how things are causally connected, because that depends on using a mental representation.
2 Because behavioral control is based on links that are triggered one after the other, actions are applied in a one-after-the-other way, with string-like organization. There can be no hierarchical embedding of programs of action.

3 Results just happen, because the past history of learning guides behavior. Results are therefore not foreseen in advance: there is no anticipated consequence (goal) when the animal starts to act.

So, we want to study the evolution of insight and mental representation abilities. Wouldn't the obvious approach be to work out which are the really cognitive bits of animal behavior and discard the rest as "just learned"? The cognitive bits would then form a lean and neat collection of data for efficient evolutionary comparison with humans. Horses for courses!

This triaging approach is dominant in much contemporary thinking about animal behavior. I think it is quite wrongheaded, and I explain why in this book. *Evolving Insight* begins (in Chapter 2) with a return to the history of human experimental psychology and the reasons for the rejection of associationism and behaviorism. I ask the question, what was found wanting, exactly? I criticize the idea that it is an empirical question whether animal behavior "is" associatively learned or cognitive. Instead, I suggest you have to decide which framework is most useful for applying science to animal behavior. I argue the cognitive framework is the right one to adopt, but I could be wrong. It might just be that in the long run, the associationist framework proves to have benefits that I cannot see right now. What isn't going to happen is a neat division between an animal's cognitive behavior and its associatively learned behavior (or between the learning sort of animal and the cognitive sort).

You can get to the same conclusion—that one should take a cognitive approach—on quite other grounds: that it allows the cross-species comparison that is essential for using animal data to reconstruct the cognitive evolution of humans. The idea of applying behaviorism and learning theory principles to human behavior has been comprehensively rejected in human experimental psychology. Given that fact alone, we need a cognitive approach to animal behavior. Since the human data for evolutionary comparisons is expressed in cognitive terms, then so must the animal data be.

After the slight digression of Chapter 2, my strategy will be to sift through the evidence of animal behavior from situations that might reveal evidence of insight. You cannot expect every chapter to end on a high note, which can be a bit disappointing but is inevitable. Building mental models of the world is not computationally the simplest way of solving problems; and, as is often pointed out, if evolution can find a simple way that works, it is likely to use it! The process of sifting evidence will work through the obvious, "sexy" topics early on, especially ones that relate to social intelligence, and sidle round to other areas that I suspect may be the key ones. The aim will be to find a workable theory of how some species of animal developed the ability to internally represent/understand how things work, and to think new things.

Since human mental superiority is most obvious in our language and speech, I will begin in Chapter 3 by looking at the closest animal system: the vocal communication of non-human primates. While interesting in many ways, primate calls turn out to differ fundamentally from language in a critical aspect, called intentionality: whereas gestural communication, at least in apes, does not. Chapter 4 will therefore consider what we can learn from the gestures of great apes. Turning more directly to how animals might gain the information needed to understand each other, Chapter 5 looks at ways in which an individual's mind is given away by its gaze, and whether using this information implies insight or not. Much of this work is based on the idea that understanding each other is a major intellectual challenge for social animals, so much so that it was this that led to the greater brain enlargement we see in primates and some other highly social species. Chapter 6 therefore considers how social "sophistication" might work, and under what circumstances does it require insight. Chapter 7 continues the theme of the effects of sociality by looking at the benefits that can be gained from tapping into others' knowledge and whether this simple sort of culture can be said to affect species' intelligence. True social insight implies some sort of "theory of mind" in order to understand the world from another's perspective, and this is the topic of Chapter 8. By this point, I hope that the reader will have come to share my own suspicion that the origins of insight—however much our having insight into others' minds may regulate our social lives—might not have been entirely a consequence of social demands. Chapter 9 uses the comparison between the highly social monkeys and the great apes to suggest that the origin of insight might be related to feeding—but feeding in smart ways. The key idea is that the origins of mental representation and computational thinking might, after all, have been driven by pressures from the physical world—acting on animals already equipped with large brains and living in richly complex societies. But if a need to feed in smart ways was the driver, what was the result? Chapter 10 examines the possibility that the result was in how physical causality is understood—but finds only weak support. However, in Chapter 11, I develop the idea that the critical advance was instead a matter of how behavior is perceived and understood. I suggest that the ability to parse behavior—an ability that initially evolved because of the need to acquire new skills, some of which were intricate and complex—led to greater things. These included a way of understanding causes, without much physics, and understanding intentions, without much mentalizing. I will also argue that even today, when we humans have the linguistic ability to understand at a deeper level of insight, we may still rely on that kind of insight that we share with great apes and a few other species of animal. Thus, a full understanding of the origins of insight in human evolution requires two different mental mechanisms.

Chapter 2

Why are animals cognitive?

The need for cognitive explanation in animal behavior

The study of animal behavior[1] has revealed many remarkable ways in which individuals deal effectively with their environment, some of which raise controversial issues of interpretation. Scrub jays, for instance, adjust their food-hiding according to the likely competition from other jays. If a competitor has seen them cache food, and they have suffered pilfering before, they re-cache in private (Emery and Clayton 2001). If privacy is denied them, they prefer to cache behind barriers, and if there are none they choose badly lit spots furthest from the competitor (Dally et al. 2004). Denied all these options, they fall back on a strategy of confusion, multiply re-caching their foods: most dramatically, they adapt this strategy to specific individuals. If one jay watched them cache in one area, and another in a different place, then they tailor their re-caching to whichever competitor appears, by re-caching just those food items that that particular jay has seen cached (Dally et al. 2006). These behaviors make good sense for a competitive forager.

In psychology, the ability to model the knowledge and beliefs of others, as distinct from one's own, is called "theory of mind" (Frith and Frith 2005). Theory of mind develops slowly in children and may be impaired in autism (Frith and Happé 2005). Because theory of mind is fundamental to linguistic communication, its attainment has often been thought a crucial step in recent human evolution, but if a bird possesses the same capacity, then our ideas about the evolution of theory of mind will need re-writing. Certainly, the scrub jay's abilities are easily described in mental theory of mind terms: the jay takes account of a competitor's viewpoint and remembers what it is likely to have seen, even keeping track of which individual jay has had the chance to find out

---

[1] This chapter is based on an opinion paper written some years ago with Lucy Bates (R. W. Byrne and L. A. Bates, "Why Are Animals Cognitive?," *Current Biology*, 16 (2006), R445–48).

about particular hiding places. It is possible, however, to interpret the bird's behavior quite differently: that it is the consequence of a complex web of associations, each association acquired according to the principles of learning theory, as derived from the laboratory study of learning in the white rat. (This sort of explanation is sometimes described as conditioning; and "behaviorism" is the philosophy that all learning is fundamentally of this nature, even in cases where we experience accompanying mental images and thoughts that seem to suggest otherwise.)

To make associative explanations work for such elaborate behavior patterns, one must take quite a bit on trust. Learning will have to be very rapid compared to that of the average laboratory rat, and sharply focused on just those specific details of the environment that turn out (afterward) to be the precise variables important in explaining how a particular behavior was learnt. Applied to the simplified environments that scientists give animals in captivity, the approach seems to work quite well. When it is extended to natural environments, crowded with distracting features that may all be salient for survival in other ways, associative learning accounts of behavior that looks insightful can sometimes appear unduly trusting! The attraction, for animal learning theorists, is the chance to de-mystify. Association learning avoids postulating mental states—there is no talk of understanding another's viewpoint, remembering what it saw, and so on. Explanation is grounded in simple phenomena, such as linking two events that often occur together in time and space, or repeating behavior that has previously led to reward. Learning theorists have shown considerable ingenuity in devising associative accounts of impressive-looking feats of animal behavior (Heyes 1993a,b). Moreover, the fact that the nervous system undoubtedly does operate on the basis of the relative strength of synaptic connections in a complex web of linkages has encouraged the idea that associative learning is the only right and proper way to understand animal behavior (Macphail 1985).

Unfortunately, associative accounts can only be tested experimentally in tightly constrained and simplified cases. Extension to the complexity of the natural worlds of animals, even that of a jay's food-caching in the laboratory, tends to be a matter of faith: reliant on *post hoc* explanation, as in the historical sciences, rather than making testable predictions, as is usually expected in natural science. Of course, psychologists who study associative learning do experimentally test predictions made by their theories, but the prediction and testing are local to the confines of highly artificial experimental situations. When associative learning is extended to account for the many complex and flexible traits observed under natural conditions, whether this is satisfactory is largely a matter of conviction and not open to verification. The problem can only get worse as explanation moves from the minutiae of single experiments and simple traits, to begin to understand the mentality of the animal as a whole. The tempting economy or "parsimony" of postulating only simple theoretical entities needs to be balanced against the power and scope of explanation over the whole range of an animal's life. The apparent simplicity of association theory can soon lead to unmanageable complexity in explaining real life.

What is needed is another "level of explanation," one that can make an interface between the massive complexity of the neural networks of the brain, and the simple efficiency of adaptive behavior in the world. This is where the cognitive level of explanation comes in: cognition offers a way of translating between theory and behavior (Morton and Frith 2004). Using the conceptual tools of cognitive science—theory of mind, working memory, focus of attention, cognitive map, number concept, counting, procedural knowledge, means-end problem-solving, and many others—allows theories to be developed, simple enough to be comprehended and used to make testable predictions in natural environments, yet tight enough to be mapped precisely onto behavior. (In principle, cognitive explanations can also be meshed with brain structures, but in animal work this is in practice more often a hope for the future, waiting for developments in imaging that can be used under relatively natural conditions.)

An everyday analogy may be made with our understanding of how a television works. There is no doubt that its "behavior" (showing moving images of things happening at places and times remote from the viewer) is fundamentally caused by the electronics. However, handing us a full circuit diagram would seldom be educationally helpful. Rather, what is needed is an intervening level of explanation, which should explain that images are sliced up, salami-fashion, and then relayed as a linear signal, traveling near-instantaneously along wires as electric potentials and across space as radio waves, finally re-assembled stripe by stripe by the electronics of the set. Only with the aid of this *cognitive model* can one start to discuss intelligently the origin of that odd flicker or annoying band on the picture, and begin to decide whether it relates to characteristics of the set, the aerial, or the transmitter. The same applies in biology. For an animal as simple as the sea-slug *Aplysia*, with fewer than 20,000 neurons and many of them large enough to experiment with individually, trying to explain behavior directly by tracing neural connections is feasible (Kandel 1979). But for vertebrates, the combinatorial explosion of possible neural interconnections in the vastly larger brain renders this an impractical task.

Cognitive models can be made to work in material other than flesh and blood. Indeed, the origins of modern cognitive neuroscience lie in the development of "intelligent machines" by Alan Turing and others. Just as a computer program can be run with solid-state electronics, or with valves and resistors, or even hydraulic components, so also, cognitive models of mental function are semi-independent of hardware. A cognitive model of some brain function, like memory or reading, can also be made to run on a computer. Relying on this freedom, some psychologists have extensively used digital computers to test their cognitive models, as simulations in which the blow-by-blow behavior of the machine and the human can be compared for a match (e.g., chess-playing, formal logic: Newell and Simon 1972; and developmental transitions in children's understanding of arithmetic: Young and O'Shea 1981). Modern psychology relies almost entirely on cognitive models of behavior. Although few are explicitly tested as simulations, clear and testable predictions can nevertheless be made from these theories, because they are expressed at the "systems level" of cognition.

The cognitive level of explanation has proved versatile for understanding human behavior, and it is no coincidence that some of the most exciting discoveries in animal behavior of recent years have begun from a cognitive perspective. As one example, consider some of the numerical abilities of animals. Chimpanzees, taught to label collections of objects with the corresponding Arabic numerals, have proved able to solve simple sums with no explicit training; and knowing numbers extends their abilities in other ways (Boysen et al. 1996). Chimpanzees cannot normally succeed in a task in which the rule is: whichever of two piles of food you point to, your companion gets to eat; you are left with the other. Trial after trial, a chimpanzee will point to the largest pile, only to be frustrated by the outcome: they simply cannot inhibit their natural attraction to the desired goal. But if Arabic numbers are substituted, they immediately solve the problem and switch to the lower number. If the test reverts to real entities, they again fail. A gray parrot, Alex, was taught various human words in order to better examine its abilities. Being a parrot, he was able to learn to speak words; but Alex understood appropriate word use rather than merely "parroting" the words (Pepperberg 1999). This enabled precise testing of Alex's understanding of number. After learning numbers as labels for quantities, he was tested with more complicated arrays of objects: for instance, several blocks and several balls, each coming in two different colors. He was reliably able to answer questions about the size of a specific subset, such as "How many green balls?," and showed that he understood verbal information of the same sort (Pepperberg and Gordon 2005). That would be an impressive ability for a 3-year-old child, and it may be that psittacine birds are innately equipped with mathematical concepts that the child only develops slowly during an extended series of interactions with adults. By questioning whether animals possess or can acquire concepts like number, set membership, and set overlap, these and other fascinating experiments have revealed hidden depths to how animals count the world.

No doubt, now that these data are known, learning theorists will manage to devise *post hoc* associative accounts to explain the animals' abilities: the point is not that any particular animal feat will "disprove" association learning. (In fact, associative explanations of realistically complex phenomena are worryingly unfalsifiable.) Rather, it is quite unclear how the topic of number and counting could ever have been explored from a standpoint of animal learning theory; and just the same applies to many other topics in animal behavior that we will visit in this book: social comprehension, spatial knowledge and navigation, imitation and teaching, and everyday understanding of physical systems such as tool use or weather. The advantage of using a cognitive level of description is that it tends to lead to interesting experiments, to novel regimes of observation, and to theories of naturally adaptive behavior that can be tested and refined.

Why does any of this need saying, if the advantages of treating animals as cognitive systems are so clear? Where's the controversy? I suspect one reason for the reluctance of biologists to embrace cognitive explanation relates to the elision—in much of the popular and even some scientific writing on animal abilities—between cognition, intelligence, and consciousness.

The tacit assumption is often made that if behavior is best understood as the result of cognitive processes, the animal is showing more intelligence than if an account in terms of association learning will suffice. The reality is more prosaic: arguing whether one animal is more intelligent than another is seldom useful, as there is no reliable scale of intellectual differences anyway. Among humans, intelligence measurements are expressed statistically in terms of a reference population's test scores, calibrated against educational achievement. An IQ of 100 is the population average; an IQ of 130 means you are two standard deviations above the mean; an IQ of 85 means you are one standard deviation below the mean, and so on. Nothing like that exists for animals, and our everyday judgments of animal "smartness" are usually based on how well an animal's social system and communicative modality mesh with our own.

Worse, the use of cognitive explanation seems to bring a whiff of consciousness along with it. For those keen to improve animal welfare, this may be manna. For anyone trained in the dry equations of animal learning it is anathema, reason enough to strive desperately for an associative counter-explanation. Both can relax: neither attitude is justified. Despite years of fascination with the possible biological function and brain localization of human consciousness, cognitive theory seldom needs consciousness in its explanation and almost never tries to explain its existence (Byrne 2000b). Although in everyday life we treat thinking as a quintessentially conscious activity, from a cognitive perspective thinking is simply a mechanistic, computational process recognizable by its products. Thinking enables the thinker to "go beyond the information given" by computing with mental representations. Animal consciousness is a fascinating area of debate, but not one likely to be resolved by empirical evidence. A cognitive approach to animal behavior has a quite different agenda: to answer "how" questions (Shettleworth 1998), by deriving and testing mechanistic theories couched in information-processing terms. The ultimate aim is to explore the variety of cognitive systems that exist in different species of animal.

Treating animals as cognitive systems is therefore not an approach that should be reserved for the most flexible and human-like species, while the behavior of simpler animals is safely explained as merely innate or as learned by association. Asking whether an animal's behavior "is" cognitive (and thus by implication "clever"), rather than associatively learnt, is not an empirical question at all. These are two different ways of studying the same behavior, and in the complex natural environments of most species only the cognitive approach leads to testable predictions. Especially relevant to the quest for the origins of insight, studying animal behavior cognitively offers the best chance of understanding the evolution of our own mind—tracing the history of cognition in primate evolution. The cognitive approach also offers the only possibility of dealing scientifically with cases of sophisticated behavioral abilities in other lineages of animal, and perhaps understanding why advanced cognition sometimes evolves. And since modern cognitive psychology phrases its findings in entirely cognitive terms, the only way animal minds can be compared to human is to express their characteristics in the same way.

Chapter 3

In the beginning was the vocalization

Vocal communication in monkeys and apes

It is a truth universally acknowledged that the greatest cognitive difference between humans and other animals lies in the use of language (I'm sure Jane Austen meant to mention this). One could go on quoting: "I think therefore I am" (but we have to trust Descartes' word for the claim that he thought); "In the beginning was the word" (John 1:1); and "Language is regarded, at least in most intellectual traditions, as the quintessential human attribute, at once evidence and source of most that is considered transcendent in us" (Wallman 1990). Words, speech, and language are so central to humanity's beliefs about humanity that it is hardly surprising that the vocal signals of other animals have always fascinated people. Surely this is where we will discover the evolutionary origins of the human capacity for insight?

The development of spectrographs in the 1950s allowed animal sounds to be examined scientifically, and affordable tape recorders encouraged researchers to document, analyze, and play back animal calls experimentally. Naturally, among the first candidates for studies were the vocalizations of apes and especially monkeys, and this early interest has led to an unrivalled set of data on vocal communication in these groups. At various times in the last 60 years, papers have been published reporting that monkey calls showed: speech-like formants; referential labels; intentional informing to others; vocal naming of addressed individuals; phonological syntax; and combinatorial changes of meaning. If all those abilities are available to today's monkeys, the same must have been true of species in our shared ancestry with monkeys over 30 million years ago (Mya). It would seem that our early *Homo* ancestors had little left to acquire, linguistically. The reality is a little different.

Vocal repertoires of non-human primates are fixed. In fact, very few mammals at all can learn or invent new sounds. Only humans, some species of cetacean, and some of the pinnipeds (seals and sea lions) have been documented to acquire new vocalizations. The contrast with birds is very striking, since most songbirds (Oscine passerines), along with hummingbirds, lyrebirds, and parrots, can all learn new sounds, even including some very un-birdlike ones. (Few will forget the superb lyrebird on the BBC's *Life on Earth* that mimicked rifle shots, car and fire alarms, and a chainsaw—presumably heard cutting down part of its forest.) Monkey calls, in contrast, are parts of a fixed system of vocal signals. In the sense in which I will use the term in this book, primate calls are "innate": not that the sounds are given at birth or completely unmodifiable by experience, but their development results from a genetically governed potential, the same vocalizations develop under a very wide range of circumstances, and the system is rather resistant to modification. All the various achievements of apes, monkeys, and prosimian primates in the vocal domain must be understood in this light: non-human primate vocal repertoires are biologically fixed. That's not to say that learning and understanding are irrelevant: individuals may learn details of *what* calls mean, *to whom* they are most effectively directed, *when* to call, or *how* to pitch calls at the most effective volumes. But non-human primates cannot acquire new calls.

Many of the most exciting findings about primate vocalizations concern what the audience can learn from hearing them, rather than why they are given. The seminal study that set off this line of research was Robert Seyfarth, Dorothy Cheney, and Peter Marler's series of playback experiments with the predator alarm calls of the vervet monkey (*Chlorocebus pygerythrus*) (Seyfarth et al. 1980a). Vervets were known to make very different sounds when they spotted predators of different sorts—leopards, martial eagles, humans, pythons, and so on—as if each call was a meaningful word (Struhsaker 1967). But it was difficult to be sure from observation alone whether hearers were discovering a predator's identity specifically from the calls: they might have spotted the predator itself, having only been alerted to danger by the call. By playing back a call when there was no predator around, the researchers were able to establish that the monkeys had genuinely gained information from the calls. Hearing a snake alarm, vervets would stand on their hind legs and scan the ground, just as if they had detected a real python; hearing an eagle alarm, they would dive into thick cover, as they would if they had seen a martial eagle overhead, and so on. Nor was this reaction just a function of the level of fear: e.g., martial eagles are much more dangerous to monkeys than pythons, so presumably more frightening. When the calls were experimentally broadcast louder or softer, the monkey reactions were made more pronounced or more muted—but they were always of the appropriate type for the specific predator. Similar results have since been found with a number of other species of monkey and prosimian, although often the calls are less specific to particular predator species.

A monkey's use of these predator calls begins in a rather non-specific way (Seyfarth et al. 1980b): baby vervets give eagle alarms to hoopoes, hornbills, and large falling leaves, as well as to martial eagles; and they give leopard alarms to buffalo, elephants, and antelope, as well as to leopards. The environmental events that trigger a specific call change during an individual's development, gradually narrowing until only the really serious dangers provoke a call. A juvenile vervet may do better than the infant, giving an eagle alarm only to large, broad-winged birds: but some of these are not dangerous, such as storks or vultures. Finally, adult vervets in Kenya only give eagle alarms to martial eagles, a major predator of mammals such as vervets. Growing monkeys apparently narrow down their usage on the basis of experience, but the broad class of referent (aerial moving object vs. quadrupedal mammal vs. long terrestrial wriggler) seems to be innate, as is the form of each signal itself. Monkeys also learn to interpret the warning calls of other species appropriately. Vervets react to the vocal alarms of superb starlings, for instance. Starlings are preyed on by a different range of species: by small hawks that do not kill vervets, but not by leopards, that do. The monkeys' reactions show they can take that into account, too. Monkeys can also combine evidence from other sources in their interpretation of warning calls. In West African forests, guineafowl give the same alarm call whether they detect hunters or leopards: for these birds, the escape strategy may be the same in either case. However, for the Diana monkeys in the trees above, very different tactics would be needed: silent departure mid-canopy to escape human hunters who have guns, noisy retreat to the highest branches to evade leopards. By "priming" the monkeys, first broadcasting sounds of people talking or leopard rasping, Klaus Zuberbühler was able to show that the monkeys used their existing knowledge as to which sort of danger was around in interpreting guineafowl alarms, which he played artificially a couple of hours later (Zuberbühler 2000a). When primed with leopard sounds, monkeys fled high and called after hearing the ambiguous bird alarms. When primed with human voices, the same alarm call caused them to flee in the mid-canopy, keeping silent.

It is evident that these monkeys can learn a great deal from hearing vocalizations given to predators by other individuals, and they are able to interpret them sensitively in context. They react only to bird calls that signal threats that they themselves share, not just blindly responding to the bird's fear. They can take account of what they already know about the local environment in interpreting ambiguous information. Does this mean that the calls are equivalent to human words? To a linguist, what is most important about words is that they can have the property of *reference*: "equivalent" would mean that the monkey calls are referential. When we use a word referentially, we deliberately intend to convey an idea in our mind to the mind of another, using a word that refers to that idea. Does this apply to monkeys? Monkeys certainly gain new information from hearing particular calls, just as we do from hearing words. Sometimes, monkeys call deliberately, in the sense that their calling is voluntary. Although some alarm calls seem to be automatic responses to danger, others are only given when other members of their social group are around, showing a so-called "audience effect." Moreover, monkeys who hear an alarm call and as a result notice the predator, often do not themselves call. But do monkeys intend to convey an idea to other monkeys? Do they even intend to change the behavior of other monkeys? Evidence that they do is worryingly sparse.

Perhaps the most remarkable suggestion of intentional communication comes from the Thomas's langur, a leaf-eating arboreal species from Sumatra. Male Thomas's langurs continue to give alarm calls until just after every female in their group has also called (Wich and de Vries 2006). It seems that the males are able to remember who has called, and only relax and stop calling when they have received evidence that every female is alert. If this were the general picture from studies of monkeys, most researchers would accept that monkey alarm calls, at least, were made with intent to inform or at least to influence their group members. Unfortunately, the finding is unique, to date. In the absence of more telling observational evidence, experiments have been devised to test directly for intentional usage. Reasoning that, if callers intended to convey new ideas, they should stay silent if it is obvious that their audience can already see the danger, Cheney and Seyfarth (1990b) set up an experiment in captive monkeys. They arranged that the mother monkey could see a veterinarian approaching ("danger," for captive monkeys), and by changing the layout they varied whether the juvenile could also see the danger, or was out of view behind a barrier. The reaction of the mother was identical in both cases: she always gave alarm calls. It may be argued that the situation of a mother sensing danger to her child is just too urgent to waste time on fine discriminations. We would perhaps not expect a human mother seeing a truck bearing down on her child to first decide whether the child has seen the truck before she screamed. All the same, considering the vast number of studies of primate vocal behavior that have been carried out in the last 50 years, the lack of evidence that calls are used with intent is rather compelling.

Underwhelmed by evidence of intentionality, primate vocalization researchers have turned to the idea that has become known as *functional reference*. Moving away from asking whether the signaler intends to change the behavior of the target audience, functional reference focuses on the recipient and what it can deduce from hearing the sounds. The audience's reaction shows what they learned from hearing a call. A wide range of information has been shown to be communicated, in monkeys alone: the species of predator detected (or perhaps the appropriate escape strategy); the individual identity and motivational state of the caller; the relative rank of a third party with whom an out-of-sight caller is interacting; the level of threat to which kin are exposed; changes of habitat (e.g., moving into an open area); detection of a neighboring group, and so on. Evidently monkeys are very good at interpreting the calls of others in a referential way.

Indeed, functional reference is often taken to be a convenient way of sidestepping the annoying and merely technical difficulty—inevitable with a species lacking speech—of proving intent. The implication is that "really" monkey vocal communication is referential; it's just hard to prove. But, as will become clear in Chapter 4, it is rather embarrassing for this interpretation how easy it has been to show intent in the gestural communication of great apes. Apes show their intention to influence the actions of a specific audience, by their own clear attentiveness to the target audience, by their sensitivity to their audience's attentional state, including choice of the appropriate gesture (silent visual, audible, or tactile), by waiting expectantly for their target audience to react, and if no reaction or an inappropriate reaction is given, by their persistence and elaboration of communicative efforts. Studies of ape gesture are able to insist on evidence of intentional usage for every single gesture analyzed, yet still have quite enough gestural communication to analyze. Similar evidence of intentional usage would be easy enough to get for human speech, if anyone seriously doubted its intentional status. So why is primate vocal communication so lacking in signs of intention, yet so obviously functionally referential in many cases? The strong implication is that most non-human primate vocal communication depends on highly skilled eavesdropping, rather than true referential communication.

That interpretation is consistent with the vocal repertoires of monkeys. Consider the vervet monkeys studied by Cheney and Seyfarth. As noted already, they give several call types that are highly specific to particular classes of predator; these calls are "discrete" (each type is vocally distinct from the others). As with all other monkey species, however, much of their vocal repertoire is "graded": there are no clear boundaries between one call and another, because all possible intermediates can occur. When vervets use these graded calls, there is no clear association with specific environmental stimuli. Functional reference is known now in a number of other monkey and prosimian species, and has also been described in several non-primates such as chickens and ground squirrels. Much of this may reflect interpretations made by hearers of vocal signals that evolved for other social purposes than giving warnings. (Vervets react appropriately to the alarm calls of starlings, but the starling alarms certainly did not evolve to benefit monkeys!) Receivers are evidently good at interpreting their auditory world, but signalers may not realize they are giving them the information. The existence of some discrete, predator-specific calls, such as vervet predator alarms, shows evolution in action. Natural selection favors signals that function, in a group of kin and allies, to enhance callers' genetical representation in the population. There is no need for callers to have any intention to influence others, nor to understand why calling is valuable: intentions and understanding may be completely absent from the monkey signaling system. This would explain why a wider range of functionally referential calls is found in a small terrestrial monkey that lives in open savannah (the vervet) than in the much larger ape species, for whom few predators are serious threats. Intriguingly, eavesdropping Diana monkeys can tell whether chimpanzee screams are being given to a leopard or simply to social aggression among themselves (Zuberbühler 2000b). Some Diana monkeys, that is: the differentiation was only made reliably by monkeys that lived in the chimpanzees' core area, suggesting that it was based on learning from experience how to read chimpanzee screams.

This is a monkey-centered view of primate vocal communication. The vocal communication of apes has been little studied. The vocalizations of apes show a high degree of grading, no discrete predator-specific calls have been identified among ape calls, and apes have limited geographical distribution compared to monkeys. All these things might suggest that the apes are a bunch of failures compared to the highly successful monkeys with their smart use of functional reference—or maybe not. Graded signals might imply that ape calls are all concerned with vague, emotion-based variation, as in human grunts and whimpers, but it is hard to be sure. It is possible that the continuum is heard as a set of separate, distinct signals by the apes themselves. Just that happens with human vowel phonemes, i.e., they are perceived categorically by human hearers, though they lie on a graded continuum of pitch. Categorical perception of their species' vocalizations has been found in several species of primate, so it could be that human hearing is giving us the wrong idea about ape sounds. The lack of highly specific signals for different predators might be because apes are unable to warn their fellows about danger, or it might be that they faced few dangers for which flight is the best strategy until the arrival of people with guns. I've watched chimpanzees enter a leopard's birthing cave and emerge with the baby leopard, killing it while the mother leopard roared ineffectually but did not emerge to fight: these are not animals that run away from much. And the present-day rarity and restricted distribution of great apes may not reflect evolutionary failure. It is almost certainly to do with direct competition from their closest relative, man. Consider the remarkable rarity of ancestor species of domestic animals, such as the wild horse (extinct), donkey (rare and restricted), dromedary (extinct), Bactrian camel (very rare in remote Mongolia), and yak (rare in remote Tibet). Superabundance of one's closest relatives is not good for long-term survival.

Could it be that apes' vocal usage will prove to be based on intent to convey information to their audiences? In a still-unpublished study, Sarah Boysen some years ago replicated the experiment about a hidden danger, but she used chimpanzees instead of monkeys. Instead of mothers and infants, she used a pair of close friends, as defined by how much time they spent together and how much they groomed each other. She examined a chimpanzee's reactions when he could see a danger approaching, again a veterinarian with a hypodermic. In each of several pairs, when one chimpanzee could see the danger approaching his friend, he called loud and long, but only if the friend was not in a position to see the danger himself. If they were both in a position to see the vet, the same chimpanzee did not bother to call. Recently, this situation has been extended with field experimentation among wild chimpanzees (Schel et al. 2013; see also Crockford et al. 2012). Anne Schel, Katie Slocombe, and their team used a realistic model viper, impersonating a danger that is very real for the members of the Sonso chimpanzee community in the Budongo Forest, Uganda. Hiding the snake among leaves, the researchers waited until a target chimpanzee was near, then pulled the string to make it move, usually securing a suitably dramatic reaction of surprise and alarm. The interesting question was what happened next, because chimpanzees travel in a loose, dispersed way, often splitting off or re-joining others. Later arrivals to the place risked not knowing about the snake, if it did not move again, and would be helped by a warning given by the ones who saw it move: did they get it? Yes: when friends or allies came on the scene, walking toward the area of the snake, chimpanzees already there gave alarm barks. This evidence shows that chimpanzees are capable of intentional communication using vocalization: they wanted a specific target group—those of their friends who were most likely ignorant of a present danger—to be alert and back off. We do not know whether the calls they used in any way signaled the nature of the threat (snake) to others who heard them, in the manner of functional reference. Nor can we be sure that the callers mentally understood that their friends were ignorant and intended to inform them. Nevertheless, this experiment comes very close to showing (true) reference in natural animal communication. Consistent with taking that interpretation, in captivity chimpanzees and other great apes have been taught to use the gestures of American Sign Language (ASL), and they certainly seemed to use the gestures referentially: for instance, using them in requests, where their subsequent reactions showed that what they requested was indeed what they wanted. It is at least arguable that such behavior would not have been possible unless the apes naturally had some concept of referential communication.

These studies suggest that there may be more to primate vocal communication—specifically that of the great apes—than generally realized. But the now very extensive literature on primate vocalization helps little toward understanding the evolutionary origins of insight. Monkeys are quick to respond to subtle implications present in their acoustic world, including the calls of conspecific and other animals, and react adaptively when unusual or dangerous circumstances are thereby signaled. These things alone do not imply representational understanding, however, and the signalers themselves show no convincing evidence of planning how to affect others' behavior in specific ways, beyond the experiments described in the last paragraph. With this background, the discovery that great ape gestures were given in a clearly intentional way has sparked much interest, and it is to this topic that we turn in Chapter 4.

Chapter 4

Gestural communication in great apes

Why "in great apes," only? It may be obvious that gestural communication in sheep and rats doesn't amount to much, apart from a small range of postures and facial expressions that have clear functions as displays; but what about monkeys? Surely their hands are as dexterous as an ape's or a person's? In fact, they're not.

This chapter's focus on great apes is not a reflection of any special hand anatomy, however. Indeed, from the relative length of thumb and fingers, we would predict that a baboon—with short fingers compared to the thumb—had greater dexterity than a chimpanzee with its "bunch of bananas" fingers and tiny thumb. But it is only in the great apes that manual dexterity has become refined. The reason is simple: the hands are controlled by the motor cortex of the brain, and the representation of the hands in the motor cortex is much more extensive in great apes, disproportionately so, even with respect to the absolute increase in brain size expected from such larger animals (Deacon 1997b). As a consequence, although some monkeys may have a fairly extensive repertoire of manual gestures and postural displays which they use communicatively (Helsler and Fischer 2007), they lack the ability to learn new gestures.

Great apes are not constrained in that way (and since little is known of the corresponding abilities in the lesser apes, the gibbons and their relatives, I will often simply refer to "apes"). The ability of apes to learn new gestures was most clearly demonstrated in several projects conducted in the 1960s and 1970s. Captive apes, often kept in human-like environments, were encouraged to learn American Sign Language (ASL) by explicit training as well as demonstration (Gardner and Gardner 1969; Gardner et al. 1989; Miles 1986; Patterson and Linden 1981). The aim of these projects was to find out whether non-human great apes could acquire something functionally close to human language, using gestures to get around the limitation we noted in the last chapter: that apes cannot learn new vocalizations. Whether the ASL projects succeeded to any extent in producing an ape with language remains highly controversial, but there is no dispute about the fact that the apes learned to use a large number of gestures that were certainly not in their natural repertoires.

Deliberately targeting specific audiences

What is more, in contrast to the difficulty of demonstrating intentionality in primate vocalizations, showing that apes have a communicative intent to inform has been easy for the case of gestures. Michael Tomasello and Josep Call have charted the deliberate and goal-directed nature of gestural communication for chimpanzees, bonobos, gorillas, and orangutans. For each species, they showed that gestures were performed for a specific target audience. Signalers typically showed what they called "response waiting": that is, they would often pause to see whether their audience was going to do what they wanted. And when that did not happen, the apes showed persistence in attaining their goal, by using more gestures, or trying some other way (Call and Tomasello 2007; Genty et al. 2009; Hobaiter and Byrne 2011b; Liebal et al. 2004a; Tanner and Byrne 1996; Tomasello et al. 1985). It would have been quite understandable if such evidence of intentional usage were only occasionally given: after all, much of our speech is given without any such clear signs, and nobody doubts its intentional basis. But in fact the evidence for intention in ape gesture is positively abundant. Recent studies have been able to restrict their analyses to only those gesture instances that occur with corroborative evidence of intentionality: individual targeting, response waiting, persistence, and elaboration. Researchers have still been left with several thousands of gestures to analyze: for instance, in a two-year study of gorillas at three zoos and one field site, insisting each gesture showed evidence of intentional use reduced an original set of 9,540 actions only to 5,250 (Genty et al. 2009).

The voluntary control and intentional delivery of ape gestures contrasts with the relatively involuntary nature of their facial expressions, and apparently the apes themselves are aware of this difference. Joanne Tanner recorded the way that a gorilla, intent on a surprise play attack, repeatedly used its hand to wipe off or cover its own telltale "play-face" expression as it crept up on the other (Tanner and Byrne 1993). The play-face is what a social psychologist would call "leakage": that is, it gives away information that the individual might prefer stayed hidden. In human communication, the hands are more likely to leak information than our "poker faces"; perhaps gorillas are less adept at controlling their faces, but even among humans deliberately put-on facial expressions are seldom very convincing.

Moreover, when great apes make gestures, they appreciate the state of attention of the specific audience they are targeting. We know this because of their choice of gesture to use. Gestures can be divided into three categories according to the modalities with which they can be detected: *silent visual* gestures, which only an audience who is already looking at the signaler can detect; *tactile* gestures, which automatically force the communication onto the audience, whether attending or not; and *audible* gestures, which can also be seen but whose sound is likely to catch the attention of an unaware audience. Studies in both chimpanzees and gorillas have shown that the choice of gesture category is adjusted to the attentional state of the target: if the target audience is looking toward the signaler, it uses a silent visual gesture; tactile gestures are used for those looking away from the signaler (Genty et al. 2009; Hobaiter and Byrne 2011b; Liebal et al. 2004b). It was initially thought that some audible gestures might function purely as "attention getters," drawing attention to the semantic information conveyed by the gesturer's demeanor, as shown by its facial expression or its next (silent visual) gesture. Surprisingly, the evidence has not supported this conjecture; indeed, delivery of audible gestures seems not to be finely adjusted to the target audience's attention (Hobaiter and Byrne 2011b). Rather than choosing an audible gesture, chimpanzees have been found to move themselves around into the visual field of their target audience (Liebal et al. 2004b). This has also been investigated experimentally, by giving the opportunity to beg a treat from a person who was either facing, side-on, or faced away from the subject, but whose head might be turned in addition, so that their face could be toward the subject even though their body was not. Chimpanzees and African elephants were found to beg preferentially from people whose face was toward them, regardless of body orientation, unless the person's body was wholly oriented away from them—perhaps that suggests unwillingness to give treats at all (Kaminski et al. 2004; Smet and Byrne 2014). To appreciate that visual signals need to be made in a place where the target will see them appears insightful, perhaps related to the large brains of apes and elephants; so it is intriguing that Caroline Ristau has shown that piping plovers, a small shore bird, show a similar ability (Ristau 1991). When a predator fails to react to the bird's "broken wing" display, designed by evolution to lure a predator toward the apparently injured adult and away from the vulnerable nest, the bird flies round into their visual field and does it again! Whether the birds really have insight into how the predator's mind works, or whether the ability is a rather primitive one, evolved long ago when birds, elephants, and apes shared a common ancestor, is still unknown.

Great apes' understanding of their audiences goes even further. Erica Cartmill carried out a study of zoo-housed orangutans which experimentally varied the level of comprehension of the audience between full, partial, and none (Cartmill and Byrne 2007). A keeper sat with two bowls at his feet, one full of a preferred food (e.g., banana) and one with a non-preferred food (e.g., cucumber). The orangutan subject gestured enthusiastically toward their favorite food. The keeper, who of course was acting as Cartmill's stooge, then handed over either the cucumber, the banana, or just half of the banana. When the keeper had, by offering the cucumber, apparently failed altogether to understand the intent behind all the gesturing, the orangutans switched strategy: they still gestured, but introduced new gesture types. But when the keeper was apparently on the right track, giving half the banana, the orangutans persisted with the same gesture types, only increasing the urgency and frequency of their gesturing. The parallel with our own behavior in a charades game was a striking one: we change tactic according to how warm or cold our team is getting in interpreting our mime. Orangutans, and probably other great apes, are capable of assessing the level of understanding of their audience, as well as their physical possibilities for detecting gestures.

Gesturing in play

There is one circumstance that provokes more gesturing—and more sequences of gesture—than any other: play (Genty et al. 2009). The fact that juvenile apes play more than adults no doubt accounts for the high frequency of gestures and gesture sequences reported in juveniles. In the rough-and-tumble play of young apes, many different sorts of gesture get used. The point is often merely to continue the interaction, at the right intensity, and with the right participant. Sequences of gesturing by both or all participants seem to be used to regulate and continually adjust the ongoing activity—rather in the way that one constantly adjusts the force and direction of the push given to a heavy wheelbarrow (Genty et al. 2009). This moment-to-moment adjustment and meshing, performed by both or all participants, has been called the "dynamic dance" by Barbara King (King 2004), who describes how the meaning of the communication is negotiated or co-constructed by the participants, not a simple property of signals given.

Although gesture is used extensively in play—and is therefore particularly obvious in zoo settings where animals have more time free to play and few urgent biological needs—this may underestimate its importance in biologically important situations. This was dramatically shown when Cat Hobaiter was able to accompany two adult male chimpanzees on consortships, a mating strategy in which a male coerces and persuades a female to come with him for several days (Hobaiter and Byrne 2012). Chimpanzees have several mating strategies, but consortship is the most risky and most successful: risky, because the pair usually have to travel away from the core of the range to avoid interruption by other males, into areas where they may encounter dangerous parties of males from another community. Male chimpanzees had been noted in zoo studies to give few gestures, but here the picture changed dramatically: two-thirds of all the gestures Hobaiter recorded from adult males in her entire study were recorded from the periods when those two males were in consortship. Most of the gestures they used were silent visual, or if audible were rather quiet ones. This makes sense: any loud noise might alert and attract unwelcome attention from competitor males or extra-community males, with violent immediate results for the pair.

Developing a repertoire

Human words are learned in a social context, making language often the most powerful indication of cultural identity. Are there cultural traditions in the gestures of great apes? In the simplest sense, the answer is certainly no: there is no evidence of local populations with a wholly different repertoire of gestures to others, or even the equivalent of dialect, with many gestures specific to particular populations. If there were such local dialects, we would find more variation between populations than within them. That has been examined in chimpanzees (Tomasello et al. 1994), bonobos (Pika 2007b), gorillas (Genty et al. 2009; Pika 2007a), and orangutans (Liebal et al. 2006): all studies concur in finding the variation between and within groups to be the same. We can therefore be sure that most ape gestures are not acquired culturally. Intriguingly, however, some of these studies report that just a few gestures do show a cultural pattern of group-specific usage: 6 out of 102 gorilla gesture types (Genty et al. 2009); 2 out of 20 bonobo gesture types (Pika 2007b); and at least 1 out of 29 orangutan gesture types (Liebal 2007). Moreover, several of the actions considered to be culturally acquired in chimpanzees were communicative gestures (Whiten et al. 1999). Identification of culture "by exclusion" of environmental and genetic explanations is controversial, as we shall see in Chapter 7, but the difficulties are least for communicative signals and perhaps we should accept that there is a small contribution of culture to the gestural repertoires of great apes.

In general, though, another explanation must be sought for the developmental origin of ape gestures. Early studies that charted repertoires of chimpanzee groups in different zoos reported massive idiosyncrasy between individuals, and even at different times of life by the same individuals (Tomasello et al. 1985; Tomasello et al. 1989; Tomasello et al. 1994). However, it depends what you mean by idiosyncrasy: here, it meant that a gesture was only recorded as used by one individual during a specific period of study time of only a few months. This is very different to the everyday sense of the word, in which an idiosyncratic gesture would have to be unique to one user, never found in any other. As we've noted already, gestures are used more by young individuals: so if a short-term observation of an adult does not record a particular gesture that was noted when that animal was younger, it does not mean that the gesture has been forgotten and has vanished from the repertoire.

Tomasello and colleagues interpreted the idiosyncrasy they found as evidence that gestures were acquired by individual learning (Tomasello et al. 1994; Tomasello and Call 1997). Specifically, they proposed that each gesture began life as a physically effective act, made in a social context: say, by individual A to individual B. Then, they suggested, over a series of interactions B begins to anticipate what A wanted—based on A's preliminary intention movements or on the first part of A's effective action sequence—and reacts "early." In turn, A would come to rely on B's anticipation, and only bother to make the intention movement or the first part of the sequence: at which point, A's action is no longer physically effective, but has become ritualized into a gesture. They called this process *ontogenetic ritualization*. The idea was based on early suggestions of Frans Plooij, who had studied mother–infant interaction of chimpanzees in the field and noticed how the actions of infants become abbreviated as the mother comes to pick up the intention more quickly; he called it "conventionalisation" (Plooij 1984). Acquisition by ontogenetic ritualization has several consequences. First, the actions that become learned as gestures do not need to be the same in each individual; and their form can be arbitrary, as long as it is related to some action originally made by the signaler in that context. This was the whole point: it can explain idiosyncrasy in gesture use, although of course the same action might be ritualized into a gesture by different individuals. Second, learning is unidirectional in a particular dyad. An individual learns each gesture by reinforcement from the other in a dyad, so it may end up learning different gestures for the same purpose with different interlocutors. Third, just because the gesture is effective for A, to B, does not mean it will work for B, to A. That would require the process of ontogenetic ritualization to occur in reverse, which may not have happened. Even if it had, it might have resulted in a different action becoming ritualized as a gesture for B. Any cases of "one-way gestures" would be highly indicative of acquisition by ontogenetic ritualization; however, two different apes are quite likely to make the same intention movements or use a similar sequence of effective actions for the same purpose, so in practice one-way gestures can never be expected to be common.

Ontogeny by ritualization from once-effective movements may be particularly apt for understanding the "iconic" gestures reported in gorillas. Joanne Tanner, in a long-term study of San Francisco Zoo gorillas, found that a few gestures had the physical form of the movement pattern desired from another gorilla (Tanner and Byrne 1996; see also Genty and Zuberbühler 2014, for the same claim made for a bonobo gesture). For instance, in sexual play, a male gestured to a female with a sweeping hand movement from in front of him to his genitals—just the path that a compliant female would take. This looks like an iconic depiction of the female's future movement (as hoped for by the male), or an indication of the location he wished her to move to. However, that movement path would also be the path of his own hand in moving a female into position, with an overshoot toward his genitals caused by the absence of a real female body, which could have been learnt by ritualization from past interactions with a female. The same male used this gesture with three different females, at different times of his life when the group composition was changed, and they all apparently understood his meaning. That's not fatal to the hypothesis: once the earliest female's reactions had reinforced the male's iconic movement as a gesture to her to take up a sexual position, it would be natural for him to repeat the same iconic action with a later partner, who might in turn begin to react appropriately. Tanner herself has argued that this was likely the case (Tanner and Byrne 1999), and suggests that gorillas have the ability to "map" their anticipated or intended actions into movement patterns, which may then develop into communicative gestures by ontogenetic ritualization, given suitably encouraging reactions from partners. If so, action mapping would confer on great apes a limited ability to mime—specifically, to mime actions that they want others to do. Anne Russon, mining a large corpus of observations of orangutan behavior from a rehabilitation camp where illegally captured orangutans are gradually encouraged back into the wild, reports just that kind of mime (Russon and Andrews 2011), used very occasionally when other efforts at communication have failed. She describes, for instance, a young orangutan picking a leaf, using it to wipe mud off his forehead while maintaining eye-to-eye contact with Russon, then handing the leaf to her as a request to help her clean his face.

In the main, however, ontogenetic ritualization is unnecessary for understanding where the repertoires of great apes come from. For me, the story begins with a study of gorilla gestures that Emilie Genty and I carried out at three European zoos and at Mbeli bai in northern Congo. Categorizing by quite fine differences, we distinguished 102 different gesture types (Genty et al. 2009). When we analyzed just one site, we found plenty of gestures that seemed to be idiosyncratic to particular individuals, as others had reported. But as we added more sites to the study, these "idiosyncratic" gestures cropped up again and again: eventually, we were left with only one idiosyncratic gesture in the entire study; and perhaps significantly that gesture was only given to a human, one of the keepers at the zoo. Many gestures were found at all sites, and the frequency of use of any gesture was a good predictor of how many sites we found it at—suggesting that the absences were not real, but merely a function of sample size. Also, we looked for, but found no trace of, "one way" gestures—apart from that one case, where the keeper did not use the gorilla's gesture back to it. Where an asymmetry in usage was found, it was always an obvious product of age or sex: for instance, infants gesture to their mothers to pick them up, not the reverse.

We were forced to conclude that the great majority of gorilla gestures are species-typical, a result of the innate potential of any individual of the species to develop the same specific movement for each gesture, without learning its form by ritualization or imitation. Tomasello and colleagues had always accepted that some great ape gestures were species-typical, simply products of the animal's biology, giving the gorilla chest-beat as an obvious example; but they considered these innate signals to be given in fixed contexts, and to lack any hallmarks of intentionality. We tried dividing the gorilla repertoire into those gestures for which an origin from intention movements seemed reasonable, as judged from gesture form, and those where it was not, which we took to be species-typical. For instance, a gorilla chest-beat deters rival males in other groups, which might be attacked and bitten or beaten if they do not get the message; but there is no obvious way in which biting or beating could turn into a chest-beat by ritualization. We then compared the usage of the "putatively ritualized" set with the definitely species-typical ones and found no difference at all. Both sets were used flexibly across several contests, with clear signs of intention to communicate, as judged by persistence, response-waiting, and their targeting a specific audience, and with usage appropriate to the audience's attentional state. The most obviously species-typical gestures were used as intentionally and as flexibly as the most easily "ritualized" ones. We concluded that almost all the gestures of the gorilla were a result of the species' innate potential to produce them. As with cultural learning, ritualization is a possible means of acquisition, but it is the exception not the rule.

It might be, of course, that gorillas were unusual in having a gestural system based almost entirely on the innate species potential; they are slightly less closely related to humans than chimpanzees and live in simpler social organizations. But when Cat Hobaiter recorded the repertoire of wild chimpanzees at Budongo, Uganda, she found the same story (Hobaiter and Byrne 2011b). By examining how the repertoire had developed over the course of her study, she established that it was close to an asymptote, where further observations would not add new gestures. At asymptote, the set of gestures recorded at Budongo were found to include almost all the gestures ever reported at any other chimpanzee field site. Admittedly, no other study had been made specifically of gesture, but gestures were noted at all of the nine long-term field sites, some with close observation of individuals for over forty years. What's more, Hobaiter found no idiosyncratic usage at all at Budongo. Every gesture she recorded was used by at least two individuals. As with the gorilla study, she compared "obvious" candidates for ritualization and clearly species-typical gestures, and likewise found no difference in intentionality, audience sensitivity, and flexibility. Like that of the gorilla, the chimpanzee repertoire is a species-typical one, based on innate potential but used in a highly intentional and flexible way—and like the gorilla's repertoire, it is very large.

These repertoires are not just species-typical, they are "family typical." When we compared the published data, including our own, on orangutans (*Pongo*), gorillas (*Gorilla*), and chimpanzees (*Pan*), we found that the "Budongo chimpanzee" repertoire included about 80 percent of the gorilla repertoire, and about 80 percent of the orangutan repertoire. As with any behavioral categorization, there are always difficulties of the best level of splitting or lumping: no two research groups ever agree entirely! Using only data from the St Andrews group, 40 gestures were the same for the African genera, *Pan* and *Gorilla* at the last count, and no less than 26 gestures proved to be the same in all three genera of great apes (see Figure 4.1). These three genera have not shared common ancestry for at least 12 million years (some would argue 15 Mya), so we are looking at a very ancient repertoire, one that

troglodytes   *paniscus*   Gorilla: 7   Pongo: 7

Pan: 7

African apes (subfamily): 40

Great apes (family): 26

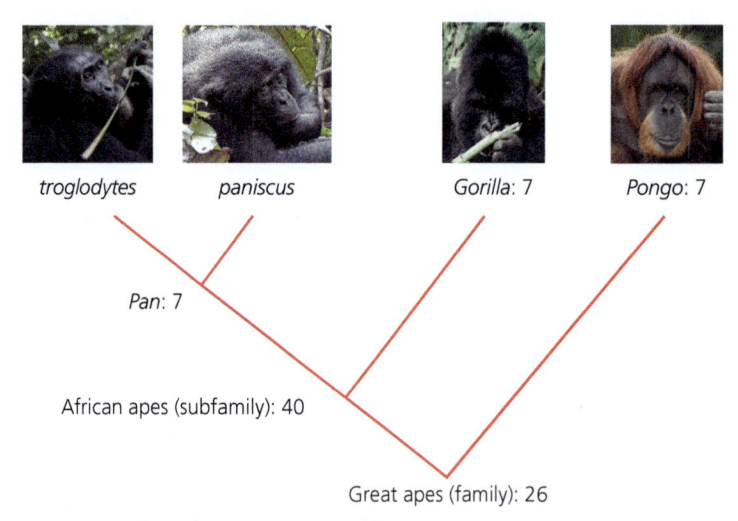

Fig. 4.1 Gestures shared among species of living great ape.

The diagram summarizes the evidence of great ape shared repertoires from the St Andrews research group, only (based on the empirical work of Erica Cartmill, Emilie Genty, Catherine Hobaiter, Lisa Orr, Kirsty Graham, Joanne Tanner, and myself). Consequently, this is very much a work in progress: we already know of cases where others have recorded a gesture for a species which we have not, so far. If they were included, the degree of sharing would be revealed to be greater, but the advantage of restricting analysis to data from a single research group is that we can ensure that each species is described using identical criteria. The numbers show the number of gestures apparently unique to each genus, on current knowledge (top row), then—moving down the diagram—those among the chimpanzees and the gorilla, and finally, those shared among all great apes. One further gesture ("object on head," a play solicitation) was recorded in only *Gorilla* and *Pongo*, not *Pan*, apparently violating the idea that species overlap results from inheritance by common descent; however, the same action has been recorded from *Pan* in play. As we gain more information on great ape repertoires, the numbers at the top of the figure are likely to diminish, and those lower down to increase. (Chimpanzee and bonobo are not separated, since the dataset for the bonobo is not fully analyzed; its inclusion would thus introduce spurious error.)

Sources: (L–R) Cat Hobaiter, Kirsty Graham, Richard Byrne, Erica Cartmill

has been building up over millennia in the ancestors of today's great apes. There are differences in gestures between species, of course; these animals have the potential to augment their inherited repertoire by ritualization, as perhaps has happened in a few cases of iconic gesture or mime, or by imitation, which may explain the small number of gestures that seem to have been learned culturally. But most of the repertoire is innate, in the sense that a genetic potential guides development into the pre-ordained form over a wide range of circumstances. This is true for most animals, but most animals do not have such large repertoires: 74 for the gorilla, a slightly different 74 for the two *Pan* species combined, and 33 for the orangutan, which has yet to be studied in the wild where the full range of communicative situations can be sampled (Cartmill and Byrne 2010). Nor, as far as we know, do most animals use their gestures in a way that shows intent to communicate and an ability to take the communicative perspective of the audiences into account, even to the point of understanding how much they have understood so far. (Caution in these dismissals is suggested by those piping plovers. But in the case of a species that shows one skill unrelated to any other in its repertoire—a one-trick pony—it is always possible that evolution has created a special solution that does not rely on any understanding.)

Intended meanings

Most animal signals are made without evidence of intentionality from the signaler. Biologists therefore ask: What is the signal's function? Function is shorthand for saying how behavior works to increase the animal's Darwinian fitness. For instance, when a chaffinch sings, the song functions to attract unattached females (who may then choose to pair with the singer, mate with him, and collaborate in making a nest and rearing their young) and to deter rival males (advertising the singer's willingness to fight if they invade his territory). It is clear how these tendencies work in increasing the chaffinch's fitness, but there is no reason to think that any of this logic is understood by the bird itself. Similarly, we may loosely speak about the song "meaning" (to hearers) that the territory is occupied, but whether rival males deduce that meaning and then leave or are simply deterred by hearing the song is irrelevant.

Since great ape gestures are made intentionally, to influence a chosen target audience, it is appropriate to ask a different question: What do the gestures mean? That is, what effect do the gesturers themselves intend to cause? Strangely, there has been little research published to date that has systematically examined meaning. When asked informally, some long-term researchers on gesture are very confident that they understand what the apes intend to achieve by each different gesture; some even put these glosses in their publications. But we must be wary of the human tendency to read rich meanings into animal signals that might not really support them. People are not, contrary to popular wisdom, particularly biased by an animal's form ("anthropomorphism"). Bob Mitchell tested this empirically, giving student subjects the same stories about animals to read (Mitchell and Hamm 1997), but changing the species: children, rabbits, otters, rats, chimpanzees. He found that people's attributions of mental states—calling the actors "kind," "guilty," "clever," "sorry"—did not vary significantly with species. But we are all prone to ascribe human-like qualities to any animals that do things that remind us of human actions.

To guard against this very human trait, it is safer to take an operational view of intended meaning. A gesture's intended meaning can be identified by seeing the response of the target audience that *satisfies* the signaler. Emilie Genty and I applied this approach to the gorilla, in the first study of what ape gestures mean (Genty et al. 2009). The approach required us to identify the target audience, as the reactions of any other individuals who happen to notice the gesture are not relevant to the signaler's intent. Fortunately, the intended target of an ape's gesturing is often readily identifiable. We judged a signaler to be satisfied if it stopped gesturing when the target audience responded, in a way that might plausibly have been wanted. (It might also stop, we thought, if the target audience attacked it, but we presumed getting reactions like that was never the idea!) Fortunately again, apes often do persist until they get a reaction, so there we had plenty of interactions to analyze. If the target had not reacted in any detectable way, then even if the signaler seemed satisfied we were able to make nothing of the interaction. Of course, that means we would have had trouble detecting a gesture whose meaning was declarative, as when we say, "That's a fish," to a child: one weakness of the approach.

The results were quite surprising. As we'd hoped, the pattern of satisfactory outcomes was characteristically different for each gesture; yet most gorilla gestures seemed "multi-functional," with signalers satisfied by several different types of outcome. Also, these outcomes overlapped considerably between gestures: it would seem that gorillas were trying to achieve rather few kinds of result, even with all those different gestures in their gestural arsenal. Erica Cartmill extended this approach to orangutans, studying the gestural communication of orangutans at several different zoos (Cartmill and Byrne 2010). Of the 64 gestures identified on structural grounds, she had enough evidence to determine the intended meaning for 40. Again, almost every gesture was used for more than one purpose, and as with the gorillas, the set of purposes to which gestures were put was much smaller than the range of gestures: to initiate an affiliative interaction (play, contact, or grooming), to request an object, to share an object or share attention to it, or to suggest moving together; and two "negative" outcomes, to cause the partner to move back or to stop an action. In both gorilla and orangutan studies, playful and serious uses of gesture were analyzed together, and play contexts dominated over all others.

There's a problem. In zoo populations—where food is not limited, predators are absent, the climate is somewhat controlled, and even whether individuals mate may be prevented or encouraged by human carers—it is not surprising that play becomes a dominant context for gestural communication. And by the very nature of play, signals given in play are not used with their real meanings. That's the whole point of play, of course, and great apes like many animals have a distinctive "play-face" that indicates that the normal significance of their displays and communicative signals can be set aside. But it is important to distinguish playful uses of gestures from serious ones, if we are to have any hope of understanding what they mean: playful and serious usage should not be muddled together in analysis. If that is done, there's a risk of obscuring the real meanings of gestures (as well as getting the impression that gestural communication is mostly about play, which we've already noted is a misleading one).

A better idea of what great apes use gestures to achieve was given by a study of a large corpus of chimpanzee gestures from the field, where Cat Hobaiter and I were able to avoid the problem by analyzing play data separately. Formalizing the idea of a reaction by the target audience which apparently satisfies the signaler, as an Apparently Satisfactory Outcome (ASO), we identified 19 ASOs in chimpanzee communication. The majority of ASOs encouraged interactions to start, but—as with the orangutan data—two were about discouraging further interaction (move away, stop that). All but 10 gestures were used for more than one ASO, but in most cases (57 of the 66 gestures) one of these meanings was play-related. When behavior from play was set aside, most gestures corresponded to one or two ASOs only, and often the "different" ASOs were actually quite similar (such as make contact, move closer, and travel with me): thus a raw count of ASOs exaggerates the real ambiguity in how a gesture should be interpreted. Two-thirds of the gestures were used toward a single desired outcome more than half the time. Importantly, a gesture's meaning, in the sense of the pattern of ASOs which it evokes, did not differ between signalers: gesture meaning is population-wide, not private to cliques or families. The simplest explanation of this result is that gesture meanings, like gesture forms, are species-typical: a young ape comes genetically equipped with the potential to produce a wide range of gestures which all other members of its species will recognize and understand to some extent, and vice versa.

Looking at it the other way round, in terms of how a particular meaning might be signaled, we found considerable variation in whether an intended meaning was signaled by a single gesture type or several gestures of apparently equivalent meaning. The degree of this "redundancy" appeared to co-vary with the potential ambiguity of the meaning itself, in a way that may help illuminate the situation. Meanings that were typically conveyed by a single gesture were often well defined and unitary: for example, initiate grooming (Big Loud Scratch). In contrast, where an intended meaning could be conveyed by several different gestures, the desired outcome was often one that required some negotiation or persuasion. For example, a request to give affiliative contact (for which several gestures are used: Embrace, Rump rub, Shake hands, Bite) does not have a form of response that is always appropriate. Exactly what the signaler wants by giving the gesture may often only become clear after some further interaction or negotiation. The subtle regulation of individual social relationships is an important part of the chimpanzee reproductive strategy, in which strong alliances are formed with related or unrelated individuals of both sexes. These relationships can impact on mating success, contributing toward individual fitness. The availability of multiple gestures for meanings involved in social negotiation may allow for particularly subtle distinctions, allowing for room to maneuver in negotiation of outcomes. Intriguingly, the gestures employed in play were mostly of the social negation type; gestures with simple, well-defined meanings don't seem to be used much in play. It may be that play is used to explore socially delicate communication: even though gesture meanings are basically species-typical, a young ape may have much to learn about the appropriateness of using gestures in particular social contexts.

Using strings of gestures

Great apes often give a series of gestures to the same audience. Does this mean that they are combining gestures in meaningful ways, to create new or compound messages, as we do with words? That question has been studied in both chimpanzees and gorillas, and for both species the answer was no. Instead, series are a consequence of two different effects, both of which are aimed at increasing the understanding by the audience of the intended message. Bouts of gestures, separated by relatively long gaps of response waiting, seem to be a result of persistence in the face of initial failure (Liebal et al. 2004a): the first gesture after a short pause is most likely to be the same as that given just before (Hobaiter and Byrne 2011a), and the pause allows assessment of whether the aim has been achieved. Human observers may treat them as a series of gestures, but to the apes themselves they are separate (failed) attempts at the same goal.

At other times, gestures are given without any pauses in which the ape could assess the outcome. These rapid-fire "sequences" are differently composed: rather than repetition of the same gesture, they show high variation in the gestures that compose them (Hobaiter and Byrne 2011a). Sequences are given much more by young apes, but using a sequence does not increase the chances of getting a successful result; success is more likely to stem from a single gesture, at any age. Older animals gain this success more readily, and that is not just because they are larger and more powerful: rather, they tend to choose the "right" gesture to use. These "right" gestures are just as efficient when used by a youngster, but the youngster is much less likely to choose them.

What is going on? Recall that a young ape is equipped by its biology with the potential to make many different gestures. The meanings of some gestures overlap, so several different gestures can be used for the same purpose. What we think is happening, when apes—largely juveniles—use gesture sequences, is a matter of self-discovery. Not all of the gestural "synonyms" for a given meaning are likely to be equally effective: we have shown that some single gestures are indeed much more effective than others. But which ones? A juvenile ape may not know. It seems that young apes use rapid-fire sequences to make it more likely that at least one of their gestures gets a result. As they get older, they are more likely to know already which gesture to use, and then they don't need to bother with a whole sequence. Presumably, they learn this relative effectiveness from noting the reactions to the gestures they initially string together in a sequence. By adulthood, sequences are used much less: but for particularly tricky problems, like the difficulty a male chimpanzee has in inducing a female to accompany him in consort away from the rest of the group, sequences are again used (Hobaiter and Byrne 2011a).

This developmental theory also accounts for the fact, noted in many studies, that the recorded active repertoire of adults is much lower than that of juveniles, which in turn is larger than that of infants. The developing ape first explores its own (potential) repertoire, actively using more and more gestures, often in sequences since it is unsure which single gesture would work best. As it gradually acquires that extra knowledge, sequence use declines and many gestures are not used at all, so adult repertoires seem impoverished. But have they forgotten those gestures they used to use and have now discarded? The study of "gestural imitation" suggests not.

Gestural imitation

When it comes to learning new skills by observation, the abilities of great apes are distinctly limited, as we shall see in Chapter 11. But in one specific type of experiment, remarkable learning ability has been reported: this is the case of so-called gestural imitation, in which apes are required to copy actions made by human demonstrators. The actions are not ordinary communicative gestures or instrumental procedures, either for the humans or the apes, but are deliberately chosen to be novel movement patterns for the ape.

The first experiment was performed with a home-reared chimpanzee, but Debbie Custance developed a more rigorous method when she replicated the results with several chimpanzees in a zoo (Custance et al. 1995). The subjects were first taught the command, "Do this," using food rewards with a training set of actions. Then she introduced novel actions, paired with the same command, and video-recorded what the subject did. Naïve coders were shown the recording and had to work out which one of the actions the chimpanzee had actually seen, on each trial. Coders were readily able to identify the right action, scoring when "blind" to the original demonstration. However, Custance noted that, although the copies matched the demonstrations, they were often rather a poor match: for instance, a two-handed covering of the ears might be copied with only one hand. Very much the same results were found with the other two genera of great ape: the orangutan (Call 2001), in which a home-reared animal was the subject; and the gorilla (Byrne and Tanner 2006), in which a zoo-housed gorilla copied actions spontaneously without reward, perhaps out of boredom in a zoo. In each case, the copies bore a clear and obvious relationship to the demonstrated actions, verified by blind scoring by naïve human coders, but they were by no means all identical to them.

These data are usually interpreted as evidence that great apes can imitate arbitrary, novel actions, but there is another possibility. With the massive repertoires of gestures with which apes are naturally equipped by their biology, the demonstrations might only be facilitating or priming gestures already in the repertoire. Remember that by adulthood, many or most of the gestures in an ape's potential gesture repertoire will have been abandoned for regular use, in favor of a relatively small set of effective gestures still in the active repertoire. When the researchers checked to make sure that their planned "novel" actions were not already in their subjects' repertoires—as they did—they would only have ruled out those in the active repertoire, not the much larger number in the ape's passive or latent repertoire. This hypothesis could explain why the "copies" were often not very accurate: because they were not copies of novel actions, but rather gestures of the individual subject which had not been used in recent years, but were brought out by the facilitation of seeing them done by the experimenter.

Plausible, perhaps: but how can we be sure? Only by possessing a near-complete repertoire for the experimental subject, based on years of painstaking observation, can the idea be tested. Remarkably, exactly that did exist for one subject. The gorilla Zura was one of the subjects of an 11-year study of gesture by Joanne Tanner at the time of the study, and she was the one who spontaneously "imitated" human actions. Zura's gestural repertoire had been recorded on videotape throughout that period. So to check, we delved into Tanner's massive long-term database on that one gorilla's gestures (Byrne and Tanner 2006). And, even though all of the demonstrated gestures had been specifically chosen by Tanner herself as "novel" to Zura, on the basis of Tanner's own experience with this gorilla group, every one of them was found to have been used before in Zura's gesturing, the evidence lurking in old data files!

To be precise, the gestures Zura had long-ago performed spontaneously were what she produced in response to the demonstrations, not new copying: explaining why she failed to exactly match several of the demonstrations which differed in details. It seems then, that the priming of rarely used items in a very extensive repertoire may explain the behavior of all the great apes that have shown gestural imitation. Their behavior is "selected" by watching the demonstration, rather than built up from what they have seen done. And that shows that the gestures explored and discarded by apes during the process of growing up are not lost, but remain in their passive gestural repertoire; presumably, the apes remain aware of the meaning of the gestures, and would recognize them if the gestures were used by others, even though they no longer use them themselves.

Summary

There is undoubtedly much more to discover about ape gesture, but what we currently know paints a puzzling picture. Apes give gestures deliberately and voluntarily, in order to influence particular target audiences, whose direction of attention they clearly appreciate and take account of in choosing which gesture to use. That shows greater insight into the process of communication than has been shown for any other non-human species. Sequences of gestures given in rapid-fire succession are prominent in the play of young apes, when the constantly changing relationships can be continually modulated and adjusted, just as one keeps changing the force applied to pushing a heavy wheelbarrow. The opportunities to "negotiate" meaning between very familiar partners may result in subtle interpretations, private to them. Nevertheless, gestures are also used in evolutionarily critical circumstances, such as the risky consort behavior of wild chimpanzees. The meanings that a signaler intends to convey by using gestures are relatively few and simple, compared to human words. Because of the large size of the repertoire, and the small number of semantic contrasts it encodes, ape gestures are redundant, with extensive overlaps in meaning.

Most surprising of all, perhaps, is gesture ontogeny. No doubt, occasionally apes do add idiosyncratic action patterns to their gestural repertoire, by the mutual conditioning within a regular dyad that has been termed ontogenetic ritualization; but this is apparently much rarer than was once thought, and in the only extensive field study of gesture (in chimpanzees) there was no evidence for it at all. No doubt, occasionally a local tradition of using a gesture may develop, unique to particular social groupings, but this again appears relatively infrequent. The great majority of gestures in the ape repertoire—and that is a very large number, compared to that of most other animals—are innate, in the sense that the potential to develop a particular gestural form and use it for a particular, restricted range of purposes is part of the species' biological inheritance. It may be brought out in any member of the species by appropriate developmental circumstances. Young individuals, apparently unsure of which gestures will be most effective for their purpose, use several equivalent gestures and thereby generate rapid-fire sequences of gestures. As they gain experience, they increasingly pick the most effective single gestures and use the "scattergun" tactic of sequence production less. Adults, therefore, use far fewer gestures than young animals. Acquisition of an adult repertoire is a process of first exploring the innate species potential for a huge number of gestures, then gradual restriction to a final (active) repertoire that is much smaller. The adults have not forgotten their full, latent repertoire of gestures. If circumstances change, this latent repertoire can once more be revealed. When gestural imitation is experimentally induced, the result is that gestures from this extensive latent repertoire are primed or facilitated; because the "imitations" are in fact part of the individual's own repertoire, the match to the demonstrations is often not perfect.

We are left with a puzzle. We know that great apes can learn novel manual gestures, as is shown most obviously in the "ape language" studies of home-reared apes. So why don't they, in the case of their everyday gestural communication?

It seems possible that great apes just don't see the point of augmenting their gestural repertoire with new expressions. If this is correct, it raises the question of whether these species understand communication in quite the way that we do. A more generous way of looking at this would be to suggest that great apes' innate repertoire is so extensive that they never reach the point at which they need to communicate something more. Either way, it suggests a lack of imagination. We might draw an analogy with a situation familiar to many infant teachers: two children learning to read and write. One, who is somewhat dyslexic but bright, has difficulty mastering the mechanics of the process, but really enjoys the reading and writing they can achieve; the other soon learns the techniques, but doesn't really see the point of reading, let alone writing, because of lack of imagination. Perhaps great apes have a general limit on their imagination, rather than a specific block on using gesture to communicate.

Nevertheless, we must remember the practical limits on what we have been able to discover, especially when making such negative conclusions. All studies of natural animal communication are limited by the logistical difficulty that a human cannot be fully part of the system. In the case of vocalization, experimental playback can simulate the signals of conspecifics or predators, which helps; but even then human involvement is restricted to information that is normally gained from out-of-sight signalers, whereas much animal vocal communication is short range. With gesture, there is no likelihood of humans even partially "becoming part of the system." There is no sign that a human mimicking an ape's communicative signals or a video display of them would be accepted as a natural interlocutor.

One critical aspect of social communication, however, can be studied much more closely: how an individual animal understands another. Information about others may be gained by following their attention, by remembering their past actions and interactions, or by using information from other circumstances and deducing likely future actions. In the following chapters, we move to examine evidence about what animals can gain from all of these sources.

Chapter 5

Understanding others

Reacting to what others see and know

In humans, noticing the eye-gaze of others is a key part of understanding what they are thinking and what they are going to do. Gaining and using this information is closely linked to having a theory of mind: early gaze-following ability in infants predicts theory of mind competence at age 4–5 (Brooks and Meltzoff 2015). Yet gaze-following is something that we do almost automatically: it is hard *not* to look up if you notice someone staring intently into the sky, a fact famously used to annoy others in practical jokes. Automaticity of that kind is a hint that gaze-following may be a "primitive" aspect of our cognition: that is, something that evolved long before our own species and has been retained because it is so valuable. As modern humans, we are able to control our gaze to a considerable extent, and we can discuss the meaning of someone else's intent gaze; but it might be that the basic machinery of gaze-following derives from periods of our ancestry when these "top down" mechanisms did not exist.

Another hint to the same effect is that our eyes seem to be designed to be followed: the white sclera makes it easy to see any mismatch between eye direction and head direction. Our closest relatives among the great apes typically have brown sclera and dark-brown eye color (Kobayashi and Kohshima 1997), making it harder to follow their eye-gaze when it diverges from head direction. The fact that we possess this derived morphological trait, one that almost offers to give away information about our intentions, is strange from an evolutionary perspective. Natural selection operates on relative advantages of individuals and is easily able to explain adaptations for manipulation of others, such as concealment of one's intentions (Dawkins and Krebs 1978; Krebs and Dawkins 1984). Our white sclera potentially benefits others at the expense of the self, which suggests it has evolved because of strong benefits to kin, accruing from interpreting gaze correctly, outweighing the inevitable costs from giving away information to competitors. To flesh out the evolutionary history of the use of gaze in the human lineage, we can use evidence from non-human relatives, especially the much-studied primates.

Gaze-following by animals

Numerous experiments have shown that other species than humans are also able and prone to follow eye-gaze. Often researchers have used themselves as the test stimuli, a procedure that introduces a possible difficulty and a possible bias. A non-human species might be able to follow conspecific gaze even if it would fail with human gaze. And the difficulty of transferring natural gaze-following ability to human gaze is likely to depend on how similar the species is to us, producing a bias in favor of our closest relatives. The data from primates show that these concerns are real.

Where researchers have required subjects to follow human eye-gaze that is divergent from head-gaze, results have generally been negative in primates (e.g., Povinelli and Eddy 1996). In contrast, experiments on following human gaze by head direction have consistently shown that great apes can follow gaze (Tomasello et al. 1999). With monkeys the results have been much less consistent, and for prosimian primates (*Strepsirrhines*) such as lemurs failure is consistently reported. On the other hand, when researchers moved to using conspecifics, even when the animals were represented only with still photographs, monkeys and lemurs proved able to follow gaze readily (Ruiz et al. 2009; Tomasello et al. 1998). (The lemur finding confirmed the evidence from observation: when engaged in social interactions, lemurs show a significant tendency to follow the other's gaze direction: Shepherd and Platt 2008.) So, given a reasonable, species-fair definition of gaze-following—something like, "following the direction that a conspecific's head points in"—we can say that all non-human primate groups possess the ability. This shows that gaze-following goes back at least to the time of shared ancestry among living primates, approximately 60 Mya, but might it be more ancient still? To answer that, we need to ask whether species other than primates can follow gaze.

One obvious candidate is the dog: several breeds have been selected for "pointing," so whether dogs can also follow gaze has long intrigued their owners. Many studies agree that dogs can follow human gaze as indicated by head-turning, but not when eye-gaze is mismatched with head direction (see Reid 2009): dogs are therefore similar to great apes at this task. But dogs have shared our lives for at least fourteen thousand years (Clutton-Brock 1995), possibly much longer, so they might have been artificially selected for this ability or simply evolved it as an adaptation for exploiting the human-commensal niche. Dogs are descended from wolves: can wolves follow human gaze? Several studies, using wolves with different degrees of human rearing (aiming to be fair to wolves in comparison to pet dogs that have lived with humans all their lives), have produced very different results even when it comes to following human pointing gestures (Udell et al. 2008; Viranyi et al. 2008), which dogs do readily, and found no strong evidence that wolves can follow gaze alone. On the other hand, goats can follow the gaze of conspecifics, though not that of humans (Kaminski et al. 2005).

In any case, some species have been found to be able to follow gaze that are much more distantly related to humans than either wolves or goats: they are birds. Many birds have long bills, so the technical difficulty of following head direction is much less than in species with more-or-less spherical heads, like many mammals. Most birds have only limited eye mobility within the head, so bill direction is a particularly accurate indicator of visual attention. For instance, bald ibis have long down-curved bills, and have proved adept at following the "gaze" of a model ibis head (Loretto et al. 2010). Being able to follow gaze potentially helps a social bird like an ibis forage efficiently, but it can help in other ways. Scrub jays, which rely on privacy to safely cache surplus food for later retrieval, have been found to react to whether or not others are able to see them (Dally et al. 2005). Although this work did not show that jays can follow gaze direction, their use of more distant or dimly lit sites for caching when observed suggests some understanding of the properties of gaze. Bee-eaters, which nest colonially in sandy banks, are reluctant to visit their nest holes if a predator is nearby. Milind Watve and colleagues examined the behavior of little bee-eaters (*Merops orientalis*) in detail, and showed that they reacted specifically to the case where the predator was able to see the nest holes, and whether or not their head was turned in the right direction, rather than simply being in view of or looking at the birds themselves (Watve et al. 2002).

Since birds and mammals last shared a common ancestor 300 million years ago, gaze-following may be an anciently evolved device; the discovery of gaze-following by a tortoise supports this hypothesis (Wilkinson et al. 2010). Gaze-following gives several benefits: it can draw attention to food found by others in social foraging; it can alert an individual to the risk of losing food to local competitors; and occasionally it can alert it to the much greater risk of attack from predators whose gaze is focused on the self. This does not mean that the species concerned actually understand these functions in the way we do. Gaze-following may be innate and automatic in many animals, including humans—to judge by how easy it is to cause chaos on a crowded pavement by standing still and staring upwards! But it makes no difference to an evolved trait whether the possessor understands it or not. A gaze-following animal's attention will be drawn to useful information in the world, regardless. To delve a little deeper into the evolution of insight, we need to ask how this information is used.

Using gaze

It appears obvious that, if the ability to follow gaze has evolved in a species, this is because it enabled members of that species to react appropriately to the information it made available: focusing attention on where to forage, on another individual that might be about to contest food, or on a sudden risk from a predator. Oddly, experiments specifically constructed to study the use of information from gaze have often given negative results, even in species that show gaze-following reliably. For instance, an animal may be presented with the task of finding a single food item that is in one of several boxes, and given a "hint": the researcher orients toward and looks fixedly at the correct box. Monkeys and even chimpanzees perform poorly at this "object choice" task, even when the researcher's gaze direction is made more salient by pointing with an arm. (Embarrassingly for primatologists, dogs do rather well at it. Selective breeding for following human intentions likely played a part in this ability, but some researchers have found that human-reared wolves are also able to follow human pointing.)

A resolution of this puzzle has come from a study of lemurs by April Ruiz and collaborators, in which gaze-following and object choice were measured in the same experiment—in contrast to most previous work on the topic, which used a separate paradigm to study each ability (Ruiz et al. 2009). Although the lemurs showed gaze-following when shown a photograph of a conspecific, looking the same way as its head direction, their behavior was variable and by no means 100 percent reliable. When it came to choosing to search the place at which the conspecific was looking (the "right" answer, since food was always located there) their ability looked even shakier, a barely significant effect. However, when Ruiz analyzed the search choices according to which way the subject had looked, the results were clear and simple. Lemurs do choose to forage at the place to which their visual attention has been drawn. The "unreliable" object choice was revealed to be mainly a matter of the variability of prior gaze-following, combined with additional variability in whether they searched in the same direction as they'd looked. It seems unlikely that lemurs are actually better at object choice than chimpanzees. More likely, if the chimpanzee data is re-analyzed for gaze-following, it will be revealed that apes and monkeys also tend to choose objects in the direction to which their gaze is drawn.

The simplest interpretation of these data is that lemurs are equipped with two innate tendencies that have evolved to increase an individual's efficiency in social foraging: gaze-following, which tends to draw visual attention to objects that others have found of interest, as shown by their orientation movements of head and body, and as we have seen is a widespread ability among animals; and *gaze-priming*. We have met the idea of priming already: activation of brain representations which then automatically affect subsequent behavior. In this case, gaze-priming activates the brain representation of the place to which the lemur's gaze has become focused: drawing the lemur's attention to that place, and making it more likely that the lemur will explore it: so if this place hides a food source, it will be found. (Of course, if instead the gaze-primed object is recognized as a predator, then anti-predator behavior will be triggered rather than foraging activities.)

Nothing discussed so far necessitates that the animals have any understanding of why others look in particular directions. To be successful both in natural foraging and in specific laboratory tasks, there is no need for insight into what is going on. But is it really the case that most animals have no insight into what happens when an individual sees something? For ourselves, the mental mechanism seems so obvious: we understand that when our attention is drawn to something, seeing leads to knowledge, and it is this knowledge that ultimately changes our subsequent behavior. Our attention may be captured automatically, but what we thereby see is available to problem-solving thought and subsequent recollection. Is it so different for other species?

Knowing about seeing

Over the past 20 years, scientists' attempts to answer that question—do other species understand about knowing?—have often been controversial, and give something of a cautionary tale about the use of "negative data": that is, the cases where animal subjects fail the task set them. The pioneering experiment was devised by Daniel Povinelli and called the "guesser–knower" paradigm (Povinelli et al. 1990). A chimpanzee was allowed to become accustomed to a task in which one and only one of four places would contain food. It also learnt that it was only allowed one try at getting the treat: not good odds. But then it was provided with hints from two unfamiliar people. One of these had been in the room when the experimenter had baited the apparatus, when the place that was baited was in plain sight for them although the chimpanzee was unable to see it. The other person was out of the room at the time: call that person the "guesser," in contrast to the "knower" who'd been in a position to see. In the test trial, both people pointed at and touched different locations: which would the chimpanzee choose? In fact, it reliably tended to choose the correct one: that pointed to by the person who'd been able to see the baiting. Since the test trials were "discriminatively rewarded" (the subject got a reward if it was right, and not otherwise), Povinelli and his colleagues realized that a quick-learning animal might simply link presence-at-baiting with correct choice: a "dumb" rule of just picking whoever was there would work fine. So they used a transfer test, in which both people were present at the baiting, but now one of them had their gaze obstructed by having a bucket placed over their head. If the chimpanzee really understood that if you cannot see what's going on you may not know much about it, then they'd avoid the advice of this person, just as if they'd been out of the room. And they did! The chimpanzee chose to follow the hint from the person who had been able to see the baiting, and it was concluded that chimpanzees understand that seeing leads to knowing. Rhesus monkeys, tested in just the same paradigm, failed (Povinelli et al. 1991). It seemed that the insight that knowledge comes from perception may be a relatively recent adaptation, only in the great ape lineage.

Flaws in these conclusions quickly became apparent. The evidence that a chimpanzee had understood the transfer test and picked the right informant was based on a block of trials. One criticism, then, was that this block of tests—since discriminatively rewarded—might also have served to train a "dumb" rule, such as never trust men with buckets over their heads (Heyes 1993b). Only the first trial of the transfer was a true test: and when those first trials were analyzed, the subjects were found equally likely to follow hints from either person (Povinelli et al. 1994). Did the chimpanzees have no clue about seeing and knowing? Or was the strangeness of the situation—a human suddenly wearing a bucket over their head—so distracting and confusing that it hardly constituted a fair test? The experiment has other flaws. Using a human as an informant is very convenient, but it means the subject has to understand the communication of another species. The problem for the animal may be a matter of cross-species incomprehension, not a failure to understand that seeing leads to knowledge. Worse, the ability to understand the communication of another species is likely to get weaker with taxonomic distance: no wonder, then, that monkeys were found to perform less well than our closest relative, the chimpanzee. Finally, even had the subjects performed brilliantly, the results might still have been suspect. The problem here is that the informants—the experimenter's collaborators who offered the hints—knew very well whether their hints were true or false. The famous case of the horse, Clever Hans, stands as an object lesson that animals may be very good at reading human demeanor (Pfungst 1911). Otto Pfungst, the psychologist who investigated the mathematical genius of the horse, concluded that when the horse's hoof-strikes approached the answer to the multiplication or subtraction sum it had been set, "certain postures and movements of the questioner" were "given involuntarily by all the persons involved and without any knowledge on their part that they were giving any such signs." He showed that when the questioner did not know the answer, or the horse could not see the questioner, the horse was at a loss. It was clear that the horse excelled not at math but human behavior-reading, and in principle, if horses are that perceptive, so might a chimpanzee be. (Although in fact there is no evidence that chimpanzees are at all good at reading human demeanor. The natural and artificial selection involved in horse domestication, operating on the natural abilities of an ungulate adapted to fast herd maneuvers, may have produced behavior-reading abilities far in excess of any primate.)

Povinelli abandoned the famous guesser–knower experiment in favor of a simpler design in which the subject merely had to beg from one of two people: one was able to see the subject, the other could not. Only if they begged from the person who could see them did they get handed a treat. Again using chimpanzees as subjects, Povinelli manipulated the line of sight of each informant in various ways (Povinelli and Eddy 1996). The person could be facing toward, away, or side-on; their heads could be in line with their body, turned to the side, or looking over the shoulder; their eyes could be open or shut; their eyes could be covered with a blindfold, or the blindfold could be over the mouth instead; their head might be covered with a bag or a bucket, which might themselves have holes cut in them; the holes could be aligned with the eyes or not. After an extensive series of experiments, the results were simple: the apes were hopeless at the task. It was not to be expected that chimpanzees, with their dark sclera, would be good at noticing which way human eyes were looking: hence they'd have problems at following gaze by eye-direction, when it did not match head-direction. But when they failed to notice and react appropriately even to whether eyes were open or shut, their problem appeared more fundamental. The only manipulation to which the subjects were sensitive was the overall body orientation, facing or away. That success looks more like an innate, evolved tendency to produce effective behavior for the chimpanzee in the absence of any insight as to what is going on, just as we saw in the case of gaze-following. The researchers concluded that, surprising as it might be to many people, chimpanzees have no understanding that knowledge comes from seeing. Non-human apes are equipped with adaptations that have the effect of improving how they gain information from the environment, but they lack any insight into why the mechanisms work.

Evidence from observational studies was at variance with that picture, however. Even before any of those experiments had been devised, analysis of records of primate tactical deception—where individuals behave in some unusual way that serves to mislead others, to their own advantage—had suggested that monkeys and apes were sensitive to the line of sight of others. For example, in several carefully observed cases, monkeys and apes maneuvered into positions in which a competitor or a predator would not be able to see them from its position. This was not just a case of "if I can't see you then you can't see me," because in some cases the two individuals were in mutual visual contact (Byrne and Whiten 1990). Hans Kummer described a female baboon maneuvering inch by inch until she was in such a position that her harem-leader male could see that she was there, but not that she was surreptitiously grooming a younger male: her hands remained out of sight behind a rock barrier. An ability to compute what is in view from another's visual perspective is not the same as understanding what they know, but it does suggest an understanding of the privileged status of seeing for subsequent action. Other rare but striking observations were consistent. Frans de Waal described how a subordinate male chimpanzee that was making sexual advances to a female, when approached by a dominant male covered his erect penis until it became detumescent, thus avoiding attack (de Waal 1982). Several observations of animal signaling that we met in Chapter 4 are also relevant. Joanne Tanner recorded a gorilla covering its lower face so that its revealing "play-face" could not be seen, in order to get close enough to another to launch a playful attack; this was not a single case, but a regular tactic in play. It appears that the gorilla not only understood the relationship between another individual's unobstructed line of sight and its own knowledge, but also realized that its own unruly body could give away information. Caroline Ristau studied a beach-living species of bird, the piping plover, which like all plovers uses a "broken wing" display to lead predators away from its nest. She found that the display was not given in a way that suggested an automatic response to danger, triggered by a fixed set of stimuli; instead, if the predator did not follow the apparently injured bird, it would fly round until it was in the predator's line of sight and try again. Milind Watve and colleagues found that little bee-eaters avoided going to their nests if a predator is in a position to see their nest, but are more likely to enter the nest if the predator is looking away from it. Note that it was the line of sight to the nest that was critical, not that to the bee-eater itself.

Each one of these observations might individually be dismissed with some "special-purpose" explanation. Together, though, they presented a strong case that great apes and many other animal species are able to connect the act of seeing with its result of knowing; but observational data are often regarded as second class in experimental psychology, and in this case they were widely ignored. Even some experiments were consistent with the observational data. It was known, for instance, that chimpanzees would search for an object of interest that an experimenter was looking at, even if the object was behind a barrier for the chimpanzee (Tomasello et al. 1999). Clearly, the chimpanzees understood the geometry of another individual's line of sight, and it is quite difficult to comprehend what use this "geometrical" ability would be to an animal that could not understand what seeing achieved.

Nevertheless, the established view remained that understanding seeing was a human-unique accomplishment (Tomasello and Call 1997)—until finally, an experiment was devised to match one of the observational data. In an experimental analogue of Kummer's observation of a baboon using a rock as a barrier, Brian Hare and colleagues positioned two competing chimpanzees on either side of an arena that had a small barrier in position (Hare et al. 2000). An experimenter dropped food items while the subordinate chimpanzee was watching; sometimes the dominant was not allowed to see the arena until this baiting was done, on other trials the dominant saw the same as the subordinate. The subordinate was released slightly in advance, and his decisions consistently showed that he understood that the dominant would only know about the food he had actually seen. Given a choice of two items, one on his own side of a barrier and the other in the open, the subordinate would reliably choose the one the dominant couldn't see; unless the dominant had also watched the baiting, in which case he'd hang back out of trouble completely. If the barrier was transparent, the effect disappeared, and the subordinate acted as if there was no barrier: the role of the barrier was in affecting vision, not as an obstacle. Chimpanzees can evidently compute what others can see, when it differs from their own perspective; do they also understand about the other's knowledge? The experiment was modified, by allowing the dominant to see the baiting of the arena, then swapping it for another, equally dominant competitor (Hare et al. 2001). If the subordinate was simply afraid because the food had been seen, but did not understand what seeing led to, it should have held back when released. It did not, but took advantage of the new, ignorant competitor to rush straight for the food behind the barrier.

Subsequent work on other species has found that the ability to understand that seeing leads to knowledge, and that those who have not seen are ignorant, is found in a range of other species. In captivity, northern ravens (*Corvus corax*) as well as scrub jays have shown evidence of understanding the meaning of gaze. Experienced birds pay attention to the presence and gaze-direction of competitors, only making food caches when they are unobserved; or, if they have no choice but to cache anyway, then choosing poorly lit areas as far away as possible from the competitor, and returning to re-cache in private at the first opportunity (Bugnyar 2002; Dally et al. 2005). In one particularly telling experiment, mentioned already in Chapter 2, a scrub jay was allowed to cache with a competitor watching, but there were two competitors, one after the other (Dally et al. 2006). When #1 was watching, the jay was only allowed to cache in one place; when #2 took over, that place was covered and the jay allowed to cache in another spot. Subsequently, when allowed to retrieve the food when one of the competitors was also present, the jay specifically chose the food that the particular competitor had seen buried, and left alone those caches that it had not seen: so jays remember who has seen what. Like chimpanzees, ravens showed in experiments that they could follow gaze around obstacles, taking the geometric viewpoint of another; they took evasive action according to whether a competitor could or could not have seen where a cache was made; and even bystanders made this differentiation, showing that it was not simply based on what the caching bird could see at the time of caching (Bugnyar et al. 2004; Bugnyar and Heinrich 2005, 2006). It is hard to explain any of these data unless the birds in some way understand that seeing leads to knowing.

Even the pioneering guesser–knower experiment has been rehabilitated, in work showing that a domestic pig can also understand the consequences of seeing (Held et al. 2001). The problem of interspecies communication was solved by using pigs as informants, one "blind" to where food had been put and the other able to see what was going on. Concern about Clever Hans effects was allayed by training all "informant" pigs to go to specific places for food, regardless of what they had actually seen: so that both blind and seeing pigs were equally enthusiastic, though equally deluded. And the training potential of test trials was negated by interspersing "probe" trials with training, so that the critical test trials were not discriminatively rewarded. Although most of the pigs failed to reach the point when they could even be tested for knowledge attribution, before they were too large for experimentation (domestic pigs are highly selected for rapid growth and body size), one of the two that could showed a reliable ability to discriminate knowledge from ignorance in another pig.

This history is an indictment of the confident claims of "species inability" that were made repeatedly in the 1990s, on the basis of the failure of a small number of individuals on a particular task that was thought by the devisers to test specifically the mental capacity they had in mind. With hindsight, it seems obvious that it is unwise to make conclusions about fundamental computational capacities on such a foundation. Failure in a laboratory task can result from a plethora of other reasons, including insufficient motivation, perceiving the task materials in a way different to the experimenters, and interpreting the task as being "about" something different. Yet time and time again, respected researchers have concluded "the chimpanzee cannot do . . . " or "animals are unable to . . . ," only to be proven wrong a few years later. In most such cases, the original conclusion conflicted with evidence from more naturalistic circumstances, but observational evidence was not considered of equal status.

Summary

We now know that a very wide range of animals is both able and inclined to follow the gaze of conspecifics. Gaze-following causes the individual's focus of attention to be directed to places, objects, and other animals that are thereby mentally highlighted, a mechanism that has been called gaze-priming. The combined tendencies of gaze-following and gaze-priming are evidently valuable, as they have the power to alert the individual to new food sources, potential competitors, and cryptic predators. These tendencies are probably automatic, and therefore might be detached from any insight into their means of operation. That may be the case for many animals, but evidence from great apes and several bird species shows that some animals are able to understand how others have access to knowledge (e.g., by computing their line of sight) and what will cause them to remain ignorant. After many years of dispute between researchers who watched animals behaving under natural circumstances and those who tested them on problem-oriented laboratory tasks, there is now concord. Some animals certainly do understand that the view from another's position is not the same as one's own and that seeing leads to knowing. They distinguish between ignorant and knowledgeable competitors, and work to keep competitors ignorant of their own prized resources.

The range of species in which these abilities are found reflects the species which have been most often chosen for study. In captivity, chimpanzees are the common choice for experiments on knowledge, but evidence from the analysis of primate tactical deception suggests that a much wider taxonomic range of monkey and ape species understand the knowledge–ignorance distinction. It is no coincidence that the great majority of deceptive tactics recorded are about concealment and the manipulation of others' attention (Whiten and Byrne 1988a). Among birds, much evidence comes from two corvid species: western scrub jays and northern ravens. Corvids and primates share a number of characteristics: they have relatively large brains; many species live socially (although in the raven only as juveniles); and they are favorite choices of researchers in animal cognition. It is not yet clear which of these characteristics, if any, is critical for finding positive evidence of insight into knowledge and ignorance. The discovery of similar abilities in little bee-eaters suggests that large brains are not critical. Nevertheless, among primates, the likelihood of a species using tactics that rely on deception is closely predicted by the size of its neocortex (Byrne and Corp 2004; we return to these issues in Chapter 6). That effect is not determined by the typical size of social group, even though it is also known that group size tends to predict neocortex size, and even though living in a larger group evidently presents more opportunities for deception to be of value. It may therefore be that all primates understand that competitors can be kept ignorant by concealment and distraction, but that larger-brained species are quicker to learn applications of this universal insight, and so their ability is more readily detected by primatologists. It is tempting to implicate social living in the understanding of ignorance and knowledge, but a simpler alternative is that solitary species demonstrate their insight less often.

The root problem is that small-brained, asocial species have been studied less: researchers have picked on ravens rather than grebes, monkeys rather than armadillos. Even the domestic pig, which is relatively small brained as a result of domestication and sometimes kept solitarily in an enclosure, is popularly thought to be unusually smart and is descended from the social living wild boar. Until species-fair tests have been carried out on a range of species less glamorous than primates and corvids, it will be impossible to rule out the possibility that their insight into knowledge and ignorance is very widespread indeed. The shared common ancestor of a raven and a chimpanzee lived about 300 million years ago and is ancestral to most land vertebrates. (And readers would be right to guess that no research has yet been carried out on fishes' understanding of knowledge and ignorance!)

Finally, if an animal understands another individual's ignorance, and knows how to manipulate the situation by concealment and distraction to keep others from knowing about certain things, exactly what does this cognitive capacity amount to? Does such knowing imply a general "theory of mind," which also includes an insight into others' beliefs and misunderstandings, and the fact that others are looking at you in the same way as you are looking at them? Before probing further into the evidence for interpersonal understanding by animals, in Chapter 8, we need to develop an account of how a species with nothing more than the abilities described in this chapter might represent the world. In the process, we will examine how much of the observed variation between animals in whether they show "smart" behavior can be attributed to quantitative variation in brain size (Chapter 6), and how much "extra help" can come from merely living in a social group (Chapter 7).

Chapter 6
Social complexity and the brain

A great deal of animal social behavior looks as if the individuals think about each other, just as we do. So, do we need elaborate experiments in order to answer the question of whether non-human animals have insight into each other's minds? Our everyday impressions may of course partly reflect the natural human trait to be "generous" in attributing abilities to animals: a weakness for a scientist. But weaknesses of that sort are usually found in sentimental pet owners, whereas in this case many researchers who have spent years studying communities of social animals in the wild, especially non-human primates, also hold generous interpretations. They may still be wrong, of course, but it is important to examine this observational evidence. What sort of case for insight in these social animals does it present? Asking that question will soon lead to others. How sophisticated must behavior become, before we really need to accept that a more human-like theory of mind underlies it? How plausible are alternative accounts of what is going on?

"Simple" explanations

What is the scientific alternative to accepting those generous attributions about animal abilities? The issue is one of simplicity in explanation, or "parsimony" as it has traditionally been called by experimental psychologists. This notion has a number of origins, including the fourteenth-century philosopher William of Ockham, and it is often known as Ockham's razor—for slicing away unnecessary explanations! The most explicit statement is perhaps that of Sir William Hamilton in the nineteenth century, who referred to the law of parsimony, "which prohibits, without a proven necessity, the multiplication of entities, powers, principles or causes; above all, the postulation of an unknown force where a known impotence can account for the phenomenon" (Hamilton 1855). This is a thoroughly sensible idea, but parsimony is sometimes applied so liberally to evidence from non-human animals that it looks rather as if it is being used to keep human superiority sacrosanct. In searching for the simplest explanation of some fascinating behavior we have recorded in nature, or of some unexpected success by subjects in experimental tests, it is important to avoid cheating. That would include assuming the animals have evolved traits, or have undergone some personal histories, that are frankly implausible: just because doing so would enable us to "explain away" inconvenient data. But if we avoid such extremes, the exercise of the law of parsimony is useful (Shettleworth 2010). For present purposes, the issue is to see how far we can go in explaining the known facts about animal social intelligence on the basis of a few traits, each of which would be so widely valuable and require so little information processing that its evolution is highly plausible. For primates, we will attempt this exercise at the end of this chapter, in the section "How does social intelligence work?" But first, let's look at the evidence: why have researchers been so impressed with the social complexity of primates?

Primate social complexity

Complexity is hard to define, and no agreed metric exists to measure it (Cochet and Byrne 2014; Sambrook and Whiten 1997). However, several factors have combined to convince scientists that monkeys and apes live in more complex societies than average non-humans. Most monkey and ape species are found in semi-permanent groups, and each individual evidently knows and remembers others in the group as individuals, behaving consistently in different ways to each (Dunbar 1988). Most animals interact with one other animal of their species at a time, for instance when mating or in head-to-head contests over resources; but in monkeys and apes, third parties often affect what happens. Hans Kummer (1967) described how a female hamadryas baboon may deliberately sit in front of her leader male and then threaten a rival, so that any threat directed back at her would necessarily also be directed to the big male. Here the third party plays a passive role, but much more often supporters intervene actively, and both short-term "coalitions" and longer-term "alliances" have been found in many species (Harcourt and deWaal 1992). The importance of third parties makes social support very important for individuals: whereas many animals use only their strength or weaponry, monkeys and apes rely far more on alliances with other group members to give them power and influence. These alliances are often strongest among close kin, but in addition, friendships between non-kin are found, often lasting over a number of years, and giving benefits to both parties. Sometimes the benefits are of different kinds for each participant, as in a barter system.

The "trade currency" used by Old World species of monkey and ape to build up alliances is social grooming, repaid in support in fights or tolerance at a prized feeding site. (That is why monkeys and apes groom each other far more than would be necessary for health purposes.) Monkeys do not spread their grooming around at random. Favored targets of grooming, apart of course from close relatives, are individuals with the power to be helpful: a male baboon, for instance, might groom unrelated females and males of higher rank than himself. Those who receive grooming are likely in the future to allow mating (if they are female) or to help in fights (if they are male). Having useful allies makes good sense and it may be that primates are unique among mammals in choosing whom to enlist among potential allies (Harcourt 1992). As well as a general correlation between an individual's grooming partners and their allies, there is a direct cause–effect relationship between the two. Monkeys are responsive to a relative's distress at all times, but a monkey who has recently been groomed by a non-relative is more responsive to an experimental playback of that individual's distress call a few minutes later (Seyfarth and Cheney 1984). And the causal direction is from grooming to future aid, as was shown by a clever experiment involving golden syrup (Hemelrijk 1994b). A researcher surreptitiously dripped a little syrup on the hair of a captive monkey. This naturally attracted others, some of whom spent time apparently grooming the monkey's back—though in reality they were simply removing and eating the sticky syrup. Nevertheless, the groomed individual supported them in later contests, evidently interpreting their grooming as an indicator of favor! Grooming is no doubt pleasant to receive, but importantly for its function as "social glue" it is also time invested in the other: and devoting time to another gives evidence of commitment that cannot be faked, because if a monkey is grooming another, he cannot be feeding (Dunbar 1992a). By means of targeted grooming, monkeys and apes build up networks of allies who can generally be relied upon for future help and support.

Sometimes, however, even the strongest relationships can be disrupted by competition. When important alliances are threatened by minor conflicts, opponents will reconcile afterward, going out of their way to groom and show affiliation to the very individuals they have recently fought: repairing the damage (Cords 1997; de Waal and van Roosmalen 1979). Reconciliation is not handed out at random, but is reserved for relationships that are important in the long term. Although much of the research has been conducted on monkeys, our closer relatives, the great apes, also apparently use grooming and reconciliation in similar ways to build up and preserve strong relationships with the right allies. Monkeys and apes live lives in which a web of influence and obligation determines most outcomes that are important to individuals.

Primate social knowledge

Primatologists use their knowledge of individual primates' kinship, friendships, and social rank to understand their societies. We now know that monkeys and apes also analyze each other this way. Social knowledge goes well beyond immediate, dyadic relationships, expanding to a wide range of third-party relationships. Losers in conflicts sometimes "redirect" their frustrated aggression to weaker third parties, in humans as well as other primates: the popular joke line is that the office boy at the bottom end of a chain of such redirection has to kick the cat. Interestingly, redirected aggression after conflicts has been found to be non-random in several monkey species (Cheney and Seyfarth 1986; Judge 1982). A victim of aggression tends to redirect their aggression specifically to the young relatives or subordinate friends of the more powerful aggressor. Uninvolved relatives of the victim may even attack relatives of the aggressor. In this monkey vendetta, the choice of victims shows that they are well aware of who are the relatives of other individuals: in some way, monkeys are able to represent the kinship of others in their social group. Verena Dasser used experiments, in which monkeys were trained to pick the appropriate picture to match a sample picture they were given, to investigate monkey social knowledge. An initial training set was given, where the rewarded response was to pick the picture of a monkey that bore a particular kin relationship to the sample; then previously untested pairs were presented to see if the monkey had understood the kin concept under test (Dasser 1988). The results showed that monkeys and apes distinguish different aspects of kinship, such as who is the mother and who is the sibling of a particular individual. Field experiments, involving playback of vocalizations (Cheney and Seyfarth 1990a; Crockford et al. 2007; Kitchen et al. 2005), have shown that monkeys are also sensitive to who is dominant to whom, who belongs in which neighboring group, and how interactions between third parties are likely to go. Monkeys and apes are not just pawns in the web of social complexity; they are also socially knowledgeable agents. Primates do not have to take part in interactions themselves to learn about the relative rank and power of others. Dalila Bovet gave monkeys the chance to see interactions between an individual they knew and a stranger, and when they later met the stranger face-to-face it was clear that they had already worked out where they stood in rank to them (Bovet and Washburn 2003).

In addition to species-wide abilities, individual primates may develop manipulative tactics that are appropriate for their particular needs, such as the use of deception to get what they want (Byrne and Whiten 1990). For instance, a female gorilla, living in a small group with a powerful male who prohibits her sexual contact with other subordinate males, may use a number of tactics to give her the freedom she desires. She may just "get left behind" so that she is out of sight of her leader male before she socializes or copulates; or she may solicit the male of her choice to follow her, and then carry out her actions with unusual quietness, for instance suppressing the copulation calls that she would normally make. Sometimes these tactics look very smart indeed. Andrew Whiten and I noticed a young baboon several times use the trick of screaming as if he had been hurt, just when he came across an adult in possession of a valued food resource. His mother ran to the scene and chased off the "aggressor," with the result that the youngster got the food. He used this tactic only when his mother was out of sight, and only against targets that were of lower rank than his mother. The difficulty with studying smart-looking tricks like this is that unlike alliance formation or reconciliation they are quite rare—unsurprisingly, since too often crying "Wolf!" simply doesn't work—so researchers tend not to publish their reports, and just write them down in their notebooks as interesting anecdotes. To get around this difficulty, we decided to survey a large number of experienced primatologists and collate their observations of deceptive tactics in primates, looking for recurrent patterns. We found that all groups of primate occasionally used deception in ways like that, although the precise tactics and how often they were used varied between species and between individuals (Whiten and Byrne 1988b).

There is therefore strong evidence that many species of non-human primate show a level of social sophistication and skill unusual for a mammal. This picture was increasingly documented in the many field studies of primates carried out in the latter half of the twentieth century, prompting a number of suggestions that the evolution of intelligence is related to social challenges.

The social intelligence theory

As long ago as the 1950s, Michael Chance and Allan Mead (1953) pointed to the extended receptivity of female primates and the conflict situations that this sets up for males. They argued that taking into account the movements of both the female and a competing male during maneuvering poses a peculiarly difficult problem. The complexity of solving this problem, they suggested, may have led to increase in the size of the neocortex in primates (they did not explicitly mention intelligence). In the early 1950s, sexual conflict was still seen as the basis of primate society (Zuckerman 1932), but subsequently this exclusive emphasis on male–male conflict as promoting intelligence has found little support, and this reduced the impact of Chance and Mead's speculations. The idea of social living as a challenge did not go away, however.

Alison Jolly (1966), studying lemurs in Madagascar, noticed that they lacked the intelligence of monkeys even though some species lived in similar-sized groups. She realized that this was inconsistent with the idea—popular at the time—that monkey-level intelligence is necessary for long-term group living, and she suggested instead that group living, arising without great need of intelligence, would subsequently tend to select for intelligence.

Most influentially of all, Nicholas Humphrey (1976) argued that monkeys and apes appear to have "surplus" intelligence for their everyday wants of feeding and ranging. Looking at the natural lives of primates, he could see little to challenge them intellectually, yet monkeys were widely accepted as being smart. Since evolution is unlikely to select for surplus capacity, he suggested that primate (and human) intelligence is an adaptation to social problem solving. Group living inevitably causes competition among individuals, he argued, yet it must be overall beneficial to each member or it would not occur. For each individual, this means that the use of social manipulation is favored, to achieve benefits for them at the expense of other group members—but without causing such disruption that the individual's membership of the group is put in jeopardy. Of course, if the victims become aware of their losses, they will be likely to adjust in order to avoid being manipulated in future. For this reason, the most useful manipulations are those in which the losers remain unaware of their loss, as in some kinds of deception, or in which there are compensatory gains, so no overall loss is perceived, as in some kinds of cooperation. The evolutionary result, of the selection pressure to engineer subtle forms of manipulation within the social group, is an increase in primate intelligence. Since the selective pressure applies to all group members, an evolutionary arms-race is set up, leading to spiraling increases in intelligence. Humphrey suggested that this kind of intelligence was not well tested by the gadgetry of psychologists' laboratories, thus explaining the many historic failures to find differences in intelligence between animals (see Macphail 1982; Warren 1973).

These various versions of the social intelligence theory are somewhat different to each other and were apparently derived independently; yet they share the feature that social complexity is given a causal role in the evolution of intelligence (see also Kummer 1982). The common ingredient is the idea that, for a monkey or an ape, to deal with the "moving targets" of social companions is tricky. In comparison, the problems posed by the physical world are seen as simpler—picking leaves and fruit to eat, traveling on when the local supply runs out, and keeping an eye out for predators. All versions are lumped together under the umbrella terms "social intelligence" or "Machiavellian intelligence" (Byrne and Whiten 1988). Why Machiavellian? Niccolo Machiavelli famously recommended politicians to use social manipulation for individual profit: "(It) is useful, for example, to appear merciful, trustworthy, humane, blameless, religious—and to be so—yet to be in such measure prepared in mind that if you need to be not so, you can and do change to the contrary" (Machiavelli 1532/1979). The most effective manipulator is therefore often the most cooperative and respected member of society. The alternative term, social intelligence, may seem to emphasize these "pro-social" traits, but ultimately the genetical benefit is a selfish one. Let's now look at the theory in a bit more detail.

Group living as a problem

Why do primates live in groups, anyway? Surely, group living exposes individuals to competition for food, mates, and other resources? That is quite true, but avoiding predation is more immediately critical for an individual than the day-to-day competition with conspecifics. Predator avoidance involves (1) learning where to avoid going, by discovering the "landscape of fear" from the occasional sight or sounds of predators, and indirectly from the reactions of others in the group (Willems and Hill 2009); and (2) reacting quickly and appropriately to alarm calls, and to any direct cues to proximity of danger, such as a predator's call or scent (Taylor et al. 1990). Avoidance can be highly effective—as long as you are living in a group, that is. If you only have direct cues to predator presence, the first cue you detect may be your last.

Group living can have many advantages for different species, but among primates group living is thought to have evolved mainly as an adaptation for dealing with predation risk (van Schaik 1983). Some benefits are conspicuous: many pairs of vigilant eyes and ears picking up information about predators from alarm calls and other individuals' reactions, and the possibility of collective defense among certain larger species. In addition, aggregation in a group pays because an individual's risk is "diluted." It was Bill Hamilton who first demonstrated the slightly counterintuitive fact that for an individual it is always safer to be with others—as many as possible—because a predator is then more likely to eat one of them instead (Hamilton 1971). That principle holds in most cases even if the predator is thereby getting an easier meal, for instance if groups are more conspicuous than single individuals. For each individual, risk halves every time group size doubles, and it is most unlikely that groups are all that much easier to find.

But living in close proximity to conspecifics certainly brings disadvantages, in terms of direct resource competition. Conspecifics are the worst kind of competitor for both food and mates. The resulting tension between aggregation and dispersal is vividly illustrated in the daily fission–fusion of baboons (*Papio papio*) living in arid Sahel habitats that they share with a range of predators (Sharman 1981). At night, aggregations of hundreds of baboons sleep together, but when foraging these groups split up into tiny parties of 3–8 individuals, efficiently re-aggregating in the heat of the day at shady drinking places by homing in on loud calls from other subgroups (Byrne 1981). This separation and re-joining of groups allows each individual to benefit from the dilution effects of "being one in a crowd" when most vulnerable to predation, while reducing competition at food patches. Multi-species groups, for instance the huge assemblies regularly found in tropical bird species, also provide predator protection by vigilance and dilution, whilst minimizing the costs of competition since different species will tend to occupy somewhat different feeding niches (Terborgh et al. 1990). Aggregations which are both temporary and multi-species have also been also noted: for instance, baboons and impala associate together, and here the different perceptual strengths of a primate and an ungulate may give additional advantage in avoiding stealth predation. As with Neotropical bird flocks, diet difference helps minimize competition, reducing the cost of association compared to the case of long-lasting groups of conspecifics.

Living in a group full time is therefore a demanding evolutionary option, as resource competition tends to disrupt group stability. In mammals, however, there are several families in which long-lasting social groups are regular and conspicuous: primates, cetaceans, carnivores, pigs, equids (horses and donkeys), and proboscideans (elephants). The key idea of social or Machiavellian intelligence theory is that an individual that can use its intelligence to acquire resources by stealth or cooperative tactics may be able to retain the benefits of group living on a more permanent basis. The logic will only apply if a social group is a semi-permanent aggregation. Temporary groupings, such as flocks of non-breeding ducks on lakes, migrating herds of ungulates, or fish schools attacking krill, are not predicted to have any such selective effect on intelligence: not all sociality is cognitively demanding. Enticingly, most reports of apparently "smart" behavior in mammals come from just those semi-permanently social species, consistent with the social intelligence proposal. Unfortunately, comparative psychology has found no acceptable "intelligence test for animals" (Warren 1973), so researchers have turned to measures of brain size to further examine the effects of sociality.

The social brain

Monkeys and apes have brains that are on average twice as large as those of a typical mammal of equivalent body size, whereas the human brain is six times as large (Jerison 1973). It was the neuroscientist Leslie Brothers who first proposed that the extraordinary size of the human brain has humble origins in the social problems of our remote primate ancestors (Brothers 1990), but much of the evidence in support of this idea has come from analysis of data from the field. Among primates, differences in brain size are primarily a function of the enlarged neocortex of some species (although neocortical enlargement is shadowed by size increases in the cerebellum, for most species). Robin Dunbar has shown that, for primate species, neocortex volume (whether measured in absolute terms, or in proportion to the size of the rest of the brain) varies with the size of the typical social group (Dunbar 1992b). By contrast, measures of environmental complexity, such as range area, day journey length, diet type, and how food is obtained, have usually turned out to be unrelated to brain size. Dunbar therefore suggested that the typical group size of a primate species is a good index of the social complexity faced by individuals, and that the size of neocortex in a species limits the size of social group in which individuals of that species can live, long term. (Because each individual needs to remain in a group for safety, adjustment of group size can only really happen by splitting or merging; over short periods, individuals must therefore put up with living in groups of sub-optimal sizes.) Over evolutionary timescales, the relationship is reversed: according to the social brain hypothesis, the increasing size and hence complexity of social groups served as the selection pressure promoting neocortical enlargement in primates.

In addition to primate findings, positive correlations have also been found between mean social group size and neocortex volume in other mammalian groups—insectivores, chiropteran bats, carnivores, cetaceans (Barton and Dunbar 1997; Dunbar 1998; Dunbar and Bever 1998). Significantly, the effect does not appear with the even-toed ungulates (Shultz and Dunbar 2006). That makes good sense, because in these species a group is a very different sort of thing than it is for a monkey or a mongoose. Most even-toed ungulates live in herds, groupings in which individuals show no clear signs of knowing each other: they react in each encounter on the basis of cues from the immediate situation. In a herd, individuals don't really have relationships with each other: they are just together. It's quite different in the case of semi-permanent group living, as is found in primates, social carnivores, toothed whales, elephants, and horses. In a semi-permanent group, differentiated relationships develop: it pays an individual to know who is kin to whom; to remember who supported them in the past; to keep track of whom they have invested grooming in, and so on. Only in species where groups are composed of individuals who have differentiated relationships with each other does group size present a significant computational load, resulting in the evolution of a larger neocortex. Moreover, in some primate species (Old World monkeys and apes), group size also correlates with the amount of time individuals spend in social grooming (Dunbar 1991, 1998), suggesting that in these species individuals need to spend more time building up networks of potential allies if they live in larger groups.

This picture makes good evolutionary logic, but it is all based on indirect evidence, of the need for intelligent behavior rather than behavior itself. Just occasionally, however, there has been an opportunity for direct assessment of animals' use of social manipulation. One such case is that of primate tactical deception. Recall that we accumulated a substantial corpus of records of tactical deception in primates, spanning all groups of primates, and that the frequencies of using deception varied systematically with species (Byrne and Whiten 1990).

Part of that variation undoubtedly reflected the fact that observer effort tends not to be uniform across species: terrestrial monkeys and apes are studied much more often, and much more closely, than arboreal species, for instance. But even when we corrected raw frequencies of deception for effort, differences remained in how much a species had been seen to use deception (Byrne and Whiten 1992). These differences turned out to be well predicted by neocortex volume, but not affected by rest-of-brain volume (Byrne and Corp 2004), showing that brain size directly affects the amount that a primate species relies on complex manipulative tactics.

These facts have led to widespread acceptance of the social intelligence theory. It is now "mainstream" that a major evolutionary stimulus to brain enlargement in social mammals has been the need for a larger neocortical area to facilitate more effective social manipulation, including the need to keep track of a complex network of social relationships (Brüne et al. 2003; Goody 1995; Seyfarth and Cheney 2002). In short, larger brains evolved in response to a need for greater social skill: increased brain size allowed more rapid learning, underlying the social sophistication that all monkeys and apes share.

Is social intelligence "domain-specific"?

If the advantages of efficient and skillful social manipulation led to enlargement of the corresponding part of the brain, the neocortex, was the outcome a brain specialized for social problem solving—but perhaps inept at other cognitive tasks? This question resonates with an old debate in psychology: is intelligence a single capacity to deal with all kinds of information efficiently, or are there different "intelligences" for different skills? That historic debate cast the question in terms of whether a single number, the value of "g," describes a person's intelligence; or is intelligence modular, so that each module might be more or less powerful (Sternberg 1985)? It must be admitted that psychology did not settle the matter for human psychometrics, largely because the discussion often degenerated into advocacy of rival statistical methods. However, it may be possible to make more progress with species intelligence, since the differences in ability may be expected to be on a larger scale than the subtle interpersonal variance within humans (Byrne 1995a).

Looking at evidence of cognitive sophistication in animals, many of the most striking cases certainly concern social understanding and social manipulation. The ability of monkeys, in particular, to notice and communicate about very subtle social nuances has been set in stark contrast to their seeming ineptitude in reacting to the physical world. Dorothy Cheney and Robert Seyfarth, impressed by the ability of the vervet moneys they studied to pick up information quickly and efficiently from alarm calls, tried simulating other cues to predators (Cheney and Seyfarth 1985). They made "python tracks" with a baseball. They hung an antelope carcass in a tree (which every tourist knows is how leopards cache their kills!). They used sound playback to simulate situations that were just plain odd, like a hippopotamus in a desert area. No reaction from the monkeys to any of this. Vervet monkeys didn't even show interest in those setups which pointed to the presence of their major predators. One might argue that the human-constructed simulations could never be perfect: maybe monkeys could tell fake python trails from real ones? Anecdotal evidence suggested not. Even when a real track was encountered, and the researchers realized that there was a huge python close by, the monkeys did not react—and walked right onto the snake! On the basis of this comparison, Cheney and Seyfarth suggested that monkeys possess a brain "module" of social intelligence, but (unlike ourselves) lack a module of environmental intelligence (Cheney and Seyfarth 1990a). But it is always difficult to make firm deductions from negative evidence—maybe it would be counter-productive for monkeys to make dramatic reactions to every sign of a predator, any more than we react dramatically when we see signs of an automobile, the main risk of death for most of us until we reach late middle-age. Monkeys do use the referential calls of other species, and use their memory of recent past events to deduce from ambiguous alarm calls which danger is actually present (see Chapter 3), bringing into question the proposal that monkeys are always poor at noticing natural history events.

On a modular theory, since the neocortex size has been shown to be a good predictor of social manipulation, then presumably the neocortex is adapted for social intelligence. But having a large neocortex has in fact been found to correlate with the frequency of showing other signs of intelligence in primates: learning from social companions, using tools, and innovation in behavioral repertoire (Reader and Laland 2001; see Chapter 7). It looks as if neocortex size, for primates at least, is best seen in terms of general intelligence, rather than a specifically social module (Deaner et al. 2006). It may still be that evolutionary selection for skill in social manipulation was what led to the development of large brains; but if so, the resulting problem-solving abilities also pay off in quite other areas.

Signs of advanced cognition also appear to be domain-general in corvids, not restricted to the food-getting and food-storing domains in which they have been most studied. For instance, partnerships in rooks and jackdaws share some of the characteristics of primate alliances (Emery et al. 2007), including reciprocal aid and post-conflict affiliation (Seed et al. 2007); and piñon jays can learn another bird's rank from third-party observation alone (Paz-y-Mino et al. 2004). And as in primates, measures of innovation frequency and the use of objects as tools—derived from meta-analysis of an extensive amateur literature—show clear relation to brain enlargement in birds (Lefebvre et al. 1997; Lefebvre et al. 2004). Unlike the case for primates, however, the evolutionary origin of corvid social skills is unlikely to have been the challenge of social living, since the telltale correlation of brain enlargement and social group size is missing (Emery and Clayton 2005). Any convergence in social skill between corvids and primates must therefore be at the level of domain-general intelligence, perhaps in each case driven by different evolutionary selective pressures. In the case of parrots, the other prominently large-brained taxon of birds, their range of abilities again suggests a domain-general intelligence, but lack of knowledge of their natural behavior prevents informed speculation on the evolutionary origin of their intellectual abilities.

Brain size and what it means

It is tempting, when confronted with all these examples in which brain size is shown to be correlated with measures of smart behavior or challenges overcome, to think of brain enlargement as a universally good thing. Life is not that simple.

In most groups of animals, such as mammals or birds, brain size shows a general tendency to correlate with body size. Allometric scaling is used to derive the brain size that would be expected, merely from being a member of the target group and of a particular size. This expected size can then be compared to the actual (measured average) brain size for the species. The pioneer on animal brain sizes, Harry Jerison, made this comparison by means of a ratio, which he called an encephalization quotient (EQ = measured/expected) by analogy with IQ (Jerison 1963, 1973). Many researchers have subsequently thought that the EQ ratio, or some other relative measure of brain size in proportion to body size, might be an appropriate index of species intelligence.

To the extent that the brain is an "on-board computer" for an animal, however, then it must follow operating principles common to any computing device—which do not include anything about the size or weight of the outer casing and the power supply. Rather, as Jerison himself realized, the number of components is much more likely to be the critical determinant of intellectual power. Or perhaps the number of components that are "spare" for computational work, supernumerary to those neurons taken up with routine tasks like controlling walking and digestion; or perhaps the richness of the interconnectedness of neurons, since multiple connections may enable more efficient computation. At present, there is no resolution as to which of these parameters has most influence, but all of them indicate that the *absolute* size of brain areas gives a better measure of their computational power, rather than evaluating brain size relative to body size.

We might expect any brain areas, which govern functions in which competition between individuals is critical to Darwinian fitness, to enlarge over evolutionary time. But enlargement carries risk. Brain tissue is one of the most energetically costly tissues of the body, with only the gut reaching the same levels of metabolic expenditure. Because of its unstable structure of lipid membranes maintained by active ion pumps, the brain requires a remorseless supply of energy, whether in "use" or not: and indeed at room temperature it begins to denature in a matter of minutes if its energy supply is cut off. Relative brain size should best be viewed as a measure of risk for an animal: the larger the brain, at a given body size, the larger the risk. Monkeys benefit from large brains, by being smart; but sloths and anteaters, whose brains are relatively small for their size, are better off in most ways.

That risk can be reduced in three ways. First, a trade-off may occur within the brain, with other brain parts reduced in compensation for one part's enlargement. With this perspective, we can see that the reduction of olfactory areas in simian primates (monkeys and apes) does not mean that olfaction was irrelevant in simian evolution: only that it was less important than vision, where massive increases in brain areas were occurring. Even with that olfaction/vision trade-off, simian primates have relatively large brains as mammals go, so brain enlargement must have been particularly valuable for them. Second, dietary change may allow energy to be gained more reliably and cheaply: for instance, it is likely that in early human evolution a shift in diet, either to meat-eating or to cooking (Wrangham 2009), allowed brain size to increase rapidly. Third, if a species' body size increases but its brain size does not need to, its risk is reduced. One factor in the selective pressures that led to increased body size in apes may have been just this sort of risk-reduction, from the pressing need to specialize in large brains.

As if all this wasn't complicated enough, it gets much worse when birds are involved. The problem is obvious. Bird brains are small, in absolute terms; indeed, the need for most species to fly puts very tight restraint on enlargement of any body part not contributing to flying. Yet their very small brains seem to work just as well, in terms of problem-solving abilities, as large ones do in mammals: corvids and parrots have very much smaller brains than apes, yet show many similar abilities. True, among birds the corvid group is one of the largest-brained families, but the brain of even the largest corvid, the northern raven, is walnut-sized; it is the same story for parrots. But are we comparing like with like? Birds have a very different brain organization to that of mammals, with which they last shared an ancestor about 300 million years ago. Identifying homologous structures in bird and mammal brains is fraught with difficulty, and researchers attempting direct comparison with mammals thus often find disorder (Healy and Rowe 2007). However, current evidence suggests that the nidopallium and other forebrain structures of the bird brain are homologous to the isocortex of mammals (Jarvis and Consortium 2005); and the nidopallium is considerably enlarged in corvids compared to other birds of similar sizes, such as pigeons. Nathan Emery has argued that this may solve the puzzle, in conjunction with the fact that corvids have a greater density of neurons—and consequently greater numbers of neurons—in these regions than other birds (Emery 2006): double that of the pigeon, for instance (Voronov et al. 1994). Whether taking all these things into account is sufficient to equalize the number of neurons, in functionally equivalent areas of the brains of corvids and apes, is not yet known.

How does social intelligence work? Managing without insight

Monkeys build up alliances over extended periods, they target their most useful allies, taking dominance and kinship into account, and they work hard to maintain their clique of friends by repairing temporary upsets. It all sounds very much like the networking one can observe at business meetings and academic conferences. When we also find that monkeys and apes use deception to attain their private ends, and monitor the out-of-sight interactions of others by interpreting what they overhear from their vocal signaling, it is hard to avoid the assumption that all this social intelligence is based on insight into the minds of others.

Greater intelligence and advanced cognition are phrases used freely in discussions of sociality and brain enlargement, but to a large extent they function as hedges, avoiding commitment to which particular cognitive mechanisms the animals in question are displaying. Without cooperative subjects under verbal instruction in a cognitive psychology laboratory, it may indeed be very hard to say for sure. In this chapter's final section I make an attempt to sketch what cognitive mechanisms are *minimally* implied by the data on social intelligence. It would be much easier to do this if we could compare specific cognitive components across species: working memory, selective attention, episodic memory, anticipatory planning, and so on. But that is in the future. Many of the behavioral indications of cognitive advance, and the brain size with which it is associated, show a continuous range of differences. Continuous variation points to the possibility that the evolutionary changes that resulted in these cognitive enhancements have also been continuous. That is, perhaps they produced "more of the same"—rather than any reorganization of cognitive architecture or the introduction of entirely new systems. Is that feasible, as a complete explanation? How far can variations between species be understood in terms of components that are very general in animals, differing only in whether they are worse or better in a given species?

A highly social animal will certainly require a far more discriminating perceptual system than will a solitary one, in order to distinguish those subtle cues that enable many different individuals to be recognized, to sustain focus and attention on key areas for picking up relevant information and avoid distraction by irrelevances, and to identify nuances of facial expression, body posture, or vocal timbre that indicate mood and disposition (Barton 1998, 2006). We know from human experience how powerful this sort of perception can be. The regular researchers in one long-running research study with whom I have collaborated can recognize individually almost all individuals of a population of 1,400 elephants without checking their notes! What we do not know is how "costly" perceptual recognition and categorization are, in terms of brain tissue. But visual processing takes up large areas of the human cortex, extending far beyond the so-called visual area of the occipital lobe, so this may have made a major contribution to brain enlargement (Barton 2006). Primates are specialized for vision, but in other highly social mammal groups different perceptual systems may dominate: for example, sonar in toothed whales and bats.

In addition to perceptual specializations, managing social living in a sophisticated way will require powerful means to register and employ knowledge. The fact that monkeys can learn about other individuals' rank and kinship from third-party observations suggests an ability to pick up social facts rapidly; and many of the instances of subtle social manipulation by deception must have required learning from rare, if not unique, past events (Byrne 1997a; Byrne and Whiten 1991). For instance, if a young baboon is once attacked for approaching an adult with food too close, it might scream in fear and this would cause its mother to rush to its defense—and coincidentally it might end up with the food as a windfall reward. Fine: no insight required. Coincidences like this are not likely to be common, but if the baboon knows the ranks of individuals relative to that of its own mother, if it takes note of its mother's presence or absence, and if it can learn from just one natural coincidence, then it is not hard to see how it could acquire a sophisticated deceptive tactic.

Very rapid learning, at least in social situations, is therefore another aspect where brain enlargement may increase the efficiency of an ability primitive to all vertebrates. And the sheer bulk of information that has to be acquired and remembered—about a substantial number of individuals, their typical interactive styles, their immediate and indirect kin relations, their dominance compared to oneself and to third parties, and the history of who can be relied upon to "network" within a social milieu made of many alliances—must all be learned from past events in order to respond appropriately in social interactions within a large monkey group (or, presumably, a higher-order dolphin alliance, or a wild dog pack). This would not be possible with a long-term memory system that was circumscribed or inefficient.

We can be sure that rather general, genetically encoded rules and a highly discriminating perceptual system, combined with rapid learning in social situations and an efficient long-term memory, are all *necessary* to explain social complexity in animals. As Sir Bob Geldof famously remarked after the first Live Aid concert: "Is that it?" Well, for most animals, I suspect yes. It is certainly tempting to describe the social maneuvering of non-human primates—reconciliation, alliance formation, tactical deception, referential communication—in human-like terms, as if the animals construed their actions as we do, planned them in advance, and understood their modus operandi. A closer look at what these social skills entail suggests otherwise (see also the work of Charlotte Hemelrijk, who has repeatedly shown that simple rules can produce the phenomena of monkey social skills without the need for insight, e.g., Hemelrijk 1994a, 1997; Hemelrijk and Bolhuis 2011).

Remembering relationships may present a considerable memory load. Suppose monkeys and apes are quicker than most other mammals at remembering socially relevant information about conspecifics—information like rank, kinship, whether they have used a certain tactic on a particular victim recently, whether it worked, and so on. (This is consistent with primates' known rapidity of learning in laboratory tasks.) Individuals will consequently have a greatly augmented database of relevant facts about their social companions. Much of the cut and thrust of choosing whom to ally with, making the alliances, remembering grievances, retaliating and choosing secondary victims in aggression, building up and repairing relationships, might then be based on rather general principles— principles which could be species-typical and easily encoded genetically as simple rules (Byrne 1995b, 2002a). All that's needed is a single tally of something we might call "goodness" for each individual. The tally should be augmented by indicators of kinship and indicators of commitment, and also by help for others rated "good." It should be reduced by any conflict with the self, or by observing conflicts with others who are rated "good." More idiosyncratic kinds of social manipulation, for instance use of deceptive tactics, must depend on learning (Byrne and Whiten 1997). If monkeys and apes have an enhanced ability to make rapid connections between social facts and environmental circumstances important for survival, they will tend to develop more elaborate and complicated tactics: ones that give the impression of real understanding, although actually acquired by normal trial-and-error learning. The great bulk of the "smart" social tactics of non-human primates and other species can therefore be understood without needing insight, provided the enlargement of mammalian neocortex confers abilities of highly discriminating perception, rapid learning in social contexts, and efficient long-term memory.

And what of the social knowledge shown in the last chapter? Surely that is consistent with an insightful understanding of mind—realizing that others know things about the world just as we do, that knowledge may be imperfect, and that what individuals believe leads them to act in the ways they do? Yes, it's consistent with insight, but that doesn't mean insight is the best explanation. Consider a simpler representation, one that might be glossed as something like "effectiveness." A predator that has a clear line of sight to your offspring and is looking in that direction is marked as effective, and therefore action needs to be taken; one that is behind a barrier or has its eyes closed is ineffective, so safer. A competitor that was not present when you hid some food is ineffective, whereas one that was staring at you is likely to be effective—although less so if you kept a distance, hid food in ill-lit areas, or had a barrier in between you and them when caching. Attributing effectiveness to predators and competitors, based on physically observable correlations, would allow an animal to behave in many of the ways that so impress us.

This answer may seem a rather unromantic one, but in fact a great deal of everyday human cognition is entirely based on sophisticated perception, rapid learning, and extensive memory—though we often prefer to glamorize our actions retrospectively, in richly intentional terms (Bargh and Chartrand 1999). Nor should it be surprising, to find that humans still rely on primitive mechanisms of social cognition. We come from a long line of social ancestors, to judge by the high levels of sociality of almost all our closer primate relatives, the apes and monkeys. In evolution, simple, primitive mechanisms do not necessarily get replaced by more advanced and sophisticated ones—even in humans. Evolution does not discard useful mechanisms just because they are "primitive"—a term that is confusingly used in everyday talk to mean overly simple and old fashioned, but which in evolutionary theory simply means inherited from earlier ancestors, in contrast with "derived" traits that have evolved de novo in the lineage.

We have not, for instance, lost our gaze-following and gaze-priming tendencies. There is every reason to think that, to the contrary, our attention is usefully drawn to events and objects in the world a hundred times a day by just the same mental mechanisms that are found in lemurs or ravens. However, mental mechanisms that are ancient in our lineage may well be less conspicuous to us because they have become "encapsulated modules" of effortless and automatic information processing, largely unavailable to conscious monitoring or control (Fodor 1983). For that reason, automatic gaze-following and gaze-priming are often more easily observed in others than in oneself. We are much more aware of the effortful struggles of our conscious thought, especially when forced to introspect on why we did something. People's post hoc theories of why they behaved in a particular way should be viewed with skepticism: without insight into the underlying brain process, we are liable to invent a plausible theory based on "more complex" mental processing. Social psychologists have documented a wide range of such cases, where people's explanations of what controlled their behavior proved to be incorrect (Nisbett and Ross 1980), usually in the direction of claiming more "advanced," "intelligent," or "complex" information processing. Humans are seldom slow to give themselves too much credit.

We do, of course, sometimes genuinely understand the behavior of others and ourselves, in a way that goes beyond the mechanisms discussed up to now. Is human insight uniquely human, then? In Chapter 8, we focus specifically on evidence that in some animals too there might be a bit more to it than that: that some differences among non-human animals may be qualitative ones, and that some social behavior may be based on insight into how it works. But before doing so, we need to complete our arsenal of devices that may help individuals look smart, without their having insight into what's going on: the automatic benefits of social living.

Chapter 7

Learning from others

Cultural intelligence?

In the last chapter, I was at pains to emphasize the difficulties that are faced by an individual that lives long term in a social group of conspecifics: the extra foraging costs of living with others. Only limited mitigation of this burden is possible, by means of fission–fusion ranging or cunning social manipulation. It is generally accepted that the main reason that primates nevertheless tend to live in groups is to reduce the risk of predation (van Schaik 1983). But for individuals of many primate species, which rely significantly on learning about the world, there is also another possible advantage to be had: from the various ways in which learning can be enhanced by exposure to the behavior or products of others, collectively called "social learning."

Social learning

Social learning is a huge topic in the study of animal behavior, and a host of different theories and proposals have been championed for how it may work in different cases, many of which are probably correct. All of them start from the high probability that some individuals in a group will possess knowledge that others lack. Even for species that are typically considered non-social, like brown bears, the cubs accompany their mother for many months and the mother has many skills the cubs do not. In what ways, then, can the ignorant benefit from being with the expert? My account here will be very brief, and phrased in everyday terms whenever possible, although some jargon is hard to avoid (for further details, see Byrne 1995a; Hoppitt and Laland 2008).

First, by traveling with an expert, an ignorant individual is *exposed* to a different range of local environments than it would encounter by haphazard ranging: specifically, it is likely to spend time in safe areas where food can be found. Useful learning is thereby made more likely to result from individual exploration. Second, some (perhaps all) social animals show a tendency, called *stimulus enhancement*, to interact more with places and objects with which they see others come into contact. This automatically disposes the ignorant observer to interact with a more favorable range of objects and sites in the environment, again boosting effective learning. Stimulus enhancement may include sensitivity to the reactions of the conspecific: if the animal is seen to show nausea or sickness, the place may instead be avoided. Third, individuals may notice another individual carrying out an action that they recognize to be in their own repertoire, and in turn be more likely to carry out that action. This is called *response facilitation*, and in combination with exposure and stimulus enhancement it will mean that the ignorant animal is likely to try out the actions it has seen another do, in the place where they saw them used, applied to the sort of object the other was also interacting with. The result is going to look very like deliberate copying, and the word "imitation" is often used; but remember that all these traits are proposed as mechanical, automatic tendencies.

Stimulus enhancement and response facilitation are cumbersome jargon terms, but both can be understood very simply as a matter of *priming* brain records (Byrne 1998). In priming, activation is automatically added to those records that match what is observed in a social context. Seeing another individual interacting with the environment primes or activates the brain records of the objects and actions involved (producing, respectively, stimulus enhancement and response facilitation). Thus an observer's attention and exploration are automatically directed toward the right thing to do. The ignorant animal may have no idea that what it is doing is in effect copying from someone who knows. Insight is not required. Nevertheless, the result is a powerful benefit for an ignorant but socially living individual, if others in the group possess greater expertise.

Notice that these mechanisms also produce conservatism: learning will, as a result of them, become channeled more toward the environments and methods that have already been found useful to group members—especially if those methods are already part of the repertoire and experience of the individuals who learn. Innovation of new skills will therefore be reduced if social learning is dominant. The resulting balance between the benefits of individual exploration (risky, but potentially innovative) and social learning (safer, but conservative), and how it changes according to environmental uncertainties, can be modeled mathematically (Laland 1996). The basic trends are clear enough, though: in rapidly changing environments, and where the risks of error are low, individual learning is a better bet than social learning. The brown rat is a species famously able to colonize new habitats and exploit commensal niches, and its individual learning abilities have kept generations of animal psychologists busy. (Even the rat resorts to social learning when it comes to feeding (Galef 1991). Rats returning to a burrow have their mouths examined for the scent or traces of food by those which have yet to feed, and whose choices are thus influenced by what others have survived eating.) Most of our closest primate relatives, in contrast, live in tropical moist forests—stable habitats (before recent human activities) in which many plants are highly toxic—and these primate species are therefore likely to rely strongly on social learning.

Innovation

With this background, it is no surprise that non-human primates have often been noted to be conservative, though the extent of this conservatism can still amaze us, as in the case of the Mahale mango. Mango trees were planted widely in Tanganyika by its German colonial masters before World War I, and some of those trees lie within the range of chimpanzee communities at Mahale, on Lake Tanganyika. Mangoes are as delicious to a chimpanzee as they are to us, as a moment's discussion with any zoo keeper will confirm, yet it took the wild chimpanzees of Mahale until 1981 before the first chimpanzee was observed to eat a mango (Takasaki 1983).

Anecdotal records of "unusual habits" are often reported in semi-popular and even academic journals. Simon Reader and colleagues used these records to compile an index of innovation frequency for a range of different primate species (Reader et al. 2011). Correcting this index for research effort, they found that the likelihood of innovation was well predicted by the species' neocortex size. In the same analysis, they included a measure of how much each species was reported to rely on social learning, and again found that it correlated with neocortex size. So, although reliance on social learning reduces the need for risky exploration, and therefore—one might think—should reduce the likelihood of showing innovation, in fact these two measures correlate among primates. This correlation is apparently driven by brain size differences: with the result, the researchers argue, that one can characterize a species' general intelligence g, as is done in human IQ testing. The chimpanzee, for instance, which for both theoretical and empirical reasons should be a conservative social learner par excellence, shows high levels of innovation as well, "because" it has a large neocortex. The suspicious quotation marks are because these correlational studies give no steer as to the direction of causality. Presumably brain differences are more likely to drive behavioral differences, at any given time. But over evolutionary timescales, whether brain size increase has been driven by environmental benefits from social learning, or from innovation, or from something quite different, remains unknown.

A prime arena for the beneficial use of innovation is tactical deception, in which misleading impressions are created in others to the advantage of the signaler. And since a large corpus of records of primate tactical deception already existed, contributed by a large number of experienced primatologists, I examined the tactics themselves for evidence of how they might have arisen (Byrne 2003a). In almost all cases, the tactic used deceptively was an action from the normal behavioral repertoire of the species: what was "innovative" was only a matter of the timing and context of use. This finding is consistent with previous analyses of tactical deception, suggesting that in most cases the deceptive usage may have been learnt without insight into its mechanism (Byrne 1997a), and does not support any idea that the innovation is a problem-oriented response to grasping the needs of the situation. Rather than necessity being the mother of invention, it would seem that in primate tactical deception good luck and a large brain are the real parents.

Species that are capable of distinguishing and remembering a lot of information about a social situation, if they experience many social situations, will "happen" to have more luck in innovating novel responses that benefit them, which are thus remembered and used again. These are, of course, just those highly social, large-brained species already discussed in the last chapter. Like social learning, innovation is a valuable benefit for a social species, but does not necessarily imply insight. The only caveat we should make, however, is that perhaps insightful innovations are too rare to help much in studying insight even in humans—where we can be sure that deliberate, problem-oriented innovation does occur—so to expect evidence of insightful innovation in non-humans is asking a lot.

Animal traditions

Heavy reliance on social learning, with the inevitable conservatism that this brings, is likely to result in transmission of specific know-how from individual to individual, producing the distinctive signs of a "tradition." These include: (1) most of the individuals in a tight-knit social group do something in the same way, whereas those in other social communities do the same thing differently, or don't do it at all; (2) the way a thing is done is somewhat stable over time, even over generations; (3) when a better way is discovered, by accident or by social learning from a new source, it spreads steadily through the population on an advancing front, like a bacterial growth on an agar plate, so that at certain times almost everyone within a bounded area does it by the new, better way whereas outside that area nobody does it that way; (4) since reliance on social learning is likely to be greatest when an important skill is unlikely to be discovered individually (Byrne 2007), traditionally transmitted traits are likely to be "smart" looking, skills that slightly surprise observers, rather than trivia. Archaeologists routinely use traditions of this kind to identify the makers of sites they are excavating (sometimes even giving them their name, as with the Beaker Folk), and to chart the flow of people across continents—or to chart the flow of knowledge, as it is often hard to know which has done the traveling, people or ideas.

There is no doubt that know-how is transmitted in this traditional way, in animals as well as people. Unglamorously, the best place to see this happening is probably the suburban garden. A well-studied case concerns the pilfering of the cream from the top of milk bottles left on the step of British houses. Bottles used to have cardboard tops: blue tits learned to open the bottles by tearing back the card in strips. This trick seems to have been discovered several times, independently in different parts of Britain, and the habit spread out from these innovation points—like that bacterial growth on the agar plate—until virtually every blue tit knew how to do it (Fisher and Hinde 1949). Then foil tops were introduced, and the whole process started again. A few innovative tits discovered that pecking downwards on the foil produced holes that showed the cream, and if the hole was enlarged enough the cream could be gained; and again the traditions spread across the land until they were commonplace everywhere. (So far, blue tits have not managed to get access to waxed-paper cartons on supermarket shelves.)

Bird tables have presented an ever-changing series of puzzles whose solutions have spread in the same, traditional way. Tits are colorful and popular species that people like to attract, and all tits feed by hanging upside down from unstable perches: hence, half-coconuts and hanging nut-baskets. At first, these were only used by the tits for which they were intended. Then, great-spotted woodpeckers started to peck out the coconuts, and greenfinches and house sparrows started to be seen hanging on "tit feeders" to eat the nuts. These relatively small behavioral innovations spread, until they were common in every garden. More extraordinary is the gradual spread in recent years of the "hanging upside-down" tactic in species that had seldom been observed hanging on anything before, such as the European robin and even the European blackbird, a kind of thrush, and the dunnock, a ground-living accentor. Presumably, these habits too will spread until they are regarded as normal. Commonplace to us nowadays is the fact that the normal nest site of the European blackbird is a garden shrub or creeper. Yet the idea of nesting in town gardens was originally quite unknown to blackbirds. Louis Lefebvre (2005) discovered the first report of this habit in 1888 in Bamberg, a small city in Bavaria. He was able to chart reports of the habit's steady spread across Europe, until today it is hard to imagine a blackbird failing to take up the "obvious" opportunities of urban gardens.

Gardens make bird traditions particularly easy to study, but there is no reason to doubt traditions are ubiquitous among social animals. Nor is there any reason to impute insight to the animals involved. Efficient social learning, as we have seen, need be based on no more than a few simple, innate "rules" or principles that enable individuals to profit from the know-how of others. Simple rules and principles, that on average produce distinct benefits, are just what evolution is expected to create.

Animal cultures

In humans, tradition is part of what we call culture, and in recent years excitement—and controversy—has arisen from suggestions that some animals "are" cultural in the human sense. This began with the chimpanzee (McGrew 1992; Whiten et al. 1999), spread to the orangutan (van Schaik et al. 2003), and even to animals as distantly related to us as whales (Rendell and Whitehead 2001) and fish (Laland et al. 2011). Culture, as the term has been used in these studies, is detected by studying the distribution of behavioral traits. Where intergroup differences are found, which cannot be explained by genetic or environmental differences, they are considered to be traditions; where more than one tradition is found in a single social group, it is said to have a culture (this is sometimes called the method of exclusion).

The logic goes like this. Invention is a rare event (or else all members of the species in question would be able individually to acquire the skills), but social learning allows knowledge to spread within the social network, within and sometimes between social communities. Spread is limited by breaks in the social network caused by natural barriers. Outside the network of privileged knowledge, individuals will remain in ignorance, or may acquire a characteristically different behavioral variant by virtue of membership of another social network in which a different technique has been invented. Thus, a patchy distribution of a tradition implies limitations on knowledge transmission: the distribution charts the local ignorance that results from environmental constraints on knowledge transmission. Support for these ideas has come from experimental studies in which behavioral choices, like a preference for opening a box by pulling rather than twisting its handle, have been shown to spread through zoo-housed groups of primates (e.g., Whiten et al. 2007).

Most of the intercommunity differences among chimpanzees and orangutans are of an all-or-nothing nature, where some communities simply do not do things that are commonly seen at another. With such "negative" evidence, there is always a slight concern that some unimagined factor—ecological or genetic— might after all explain the distributional pattern. There was therefore some excitement when a stylistic difference in tool making and use was found to vary between communities: in how chimpanzees consume driver ants. At Gombe in Tanzania, a long, rigid wand is dipped into the moving columns of these ferocious ants; individuals of the soldier caste swarm up the wand. Just as the first are about to reach the chimpanzee's hand, it holds up the wand and sweeps the mass of ants off in a single movement with the other hand, quickly stuffing the lot into its mouth and chewing fast to minimize the pain! At Taï in Ivory Coast, a more leisurely, one-handed method is used. A lighter, shorter stick, often with some branching at the tip, is dipped into the ant column and then swung round to allow the chimpanzee to bite off any clinging ants. This method obtains ants at only a quarter of the rate of the East African one. And local differences in efficiency are one of the hallmarks of a traditionally transmitted skill: the method you pick up depends on where you happen to live, so it may not be the most efficient. Despite the dramatic difference in efficiency, behavior at Taï has not converged on the optimal Gombe form but continues to conform to the local version. Researchers concluded that "Taï chimpanzees restrict themselves to the suboptimal solution that must be maintained by a social norm" (Boesch 1996), and "it is difficult to see how such behavior patterns could be perpetuated by social learning processes simpler than imitation" (Whiten et al. 1999). Indeed, they postulated that the more complex technique derived culturally from the simpler method by "differentiation in concert with diffusion, a process more deserving of the term 'cultural evolution'" (Whiten et al. 2001). The ant-dipping evidence was cast-iron proof that intergroup differences in behavior showed culture: ant dipping must have been invented at most twice, and different ways of achieving the same outcome became stable in different knowledge networks.

Until, that is, Tatania Humle, working at Bossou in Guinea (Humle and Matsuzawa 2002), discovered that there the same individual chimpanzees use both techniques: oops! Evidently the one-handed method cannot be universally inferior, after all; and indeed, Humle discovered that the two methods are each appropriate for different species of ant. (Until this point, primatologists had not realized there were several species of driver ant living at some sites.) With the particularly aggressive species, also found at Gombe, the two-handed sweeping method was preferred. With the milder species, especially when moving in a column rather than bivouacked as a mass, the same individual chimpanzees revert to simpler, one-handed dipping: a matter of "horses for courses," after all.

Since then, confidence in the exclusion method for identifying animal culture has sagged—perhaps unsurprisingly, since with a cognitively sophisticated behavior like chimpanzee termite-fishing, excluding alternative explanations in wild populations is tricky. True, the method of identifying a local culture by exclusion is routinely used in anthropology, where it is called ethnography, but economic and ecological constraints on humans can be readily detected by the anthropologists, just because they too are human. In contrast, the behavioral ecology of great apes is imperfectly known. The chimpanzee and the orangutan are unusual in the richness of their known diet sets and behavioral repertoires, even for primates living in tropical forests. Yet neither diet has been subject to phytochemical and nutritional examination in the kind of detail needed to reveal subtle interactions among alternative diet items; nor has there been any study of the mechanical properties of materials that may potentially be employed in tool use. This level of ignorance is unsettling, in the context of a method of analysis that depends on exclusion of ecological factors. It seems rash to dismiss the environment as a contributor to a pattern of differences, when so little is known about it.

In the case of the chimpanzee, the first species for which culture was suggested, 39 traits were considered to show variation which could not be explained by environmental constraints. Of those 39 traits, 18 involve feeding on specific plants or animals, 21 employ specific plant material as the means, and 2 involve removal of specific noxious insects: taking these together, an ecological explanation for the intergroup variation is possible in 32 of 39 cases (Byrne 2007). This worry isn't just pedantry: ecological differences in chimpanzee foods can be very subtle. Mound-building termites are found at three study sites on the eastern shores of Lake Tanganyika, yet chimpanzees at one of them (Kasoje) do not use stems to fish for them. Collins and McGrew (1987) carried out a detailed study of the termites and found that three different termite species of two genera were involved. They concluded that the chimpanzees' behavior matched the termite species available rather than reflecting cultural differences, but without their painstaking study of the insects concerned, the difference might easily have been attributed to culture. Moreover, the distributional pattern of most of the traits that vary between groups looks nothing like the spread of bacteria across an agar plate, as would be expected if the variations were caused by blocks on knowledge transmission. Rather, different ways of doing things are spottily distributed, scattered across the chimpanzee range in Africa. It would seem that many of these traits are not too difficult for chimpanzees to invent, and that invention has occurred independently at many sites. This worry is reinforced by inter-species comparison. A number of "chimpanzee cultural variants" have even been identified in the bonobo (Hohmann and Fruth 2003). These include the grooming handclasp, aimed throwing, fly-whisking with vegetation, and moss sponging (Goodall 1986; Hobaiter et al. 2014; McGrew and Tutin 1978). The bonobo is not only a different species, but also one separated from the chimpanzee by Africa's largest river. Both species must have a propensity to invent and re-invent the same actions under similar circumstances.

Doubts about the method of exclusion are now backed by empirical data. Kat Koops and colleagues (2014) have shown that many intergroup differences are in fact driven by ecology, rather than physical barriers to the spread of knowledge.

They review recent analyses of data on tool using from chimpanzees, orangutans, and capuchin monkeys and show that, in all cases, whether and how frequently individuals use tools depends on the opportunities that they have to encounter the resources for which tools are necessary and the raw materials for making tools. (In the course of these analyses, the idea that tool use is a function of ecologically sparse, arid habitats was comprehensively rejected.) Ecology cannot be "excluded"—it is what drives the behavioral variation that originally fascinated researchers on great apes.

A very different understanding of ape culture results from this reappraisal. The differences between populations are now seen to be a result of differential opportunity, rather than obstacles to knowledge transfer. This makes excellent sense. Although fragmentation of natural habitats in Africa and the Sunda islands of Sumatra and Borneo has isolated many populations of chimpanzees and orangutans, respectively, that has only recently been the case. There is no evidence that any of the natural habits of these species, even the most subtle and elaborate tool making and using, are changing rapidly: rather the reverse. Archaeological study of stone hammer and anvil use for nut cracking in the Taï forest, Ivory Coast, has shown repeated use of chimpanzee anvil sites over thousands of years (Mercader et al. 2007). Transfer of individuals between home ranges and communities, over hundreds of years, would inevitably spread the knowledge of any skill worth learning (Byrne 2007). With recently invented, fast-changing "fads" of behavior, patchiness of knowledge is to be expected, especially from the delays in transmission that would be caused by the reduced likelihood of transfer across rivers. The ape culture work, however, has not focused on fads (but see Perry et al. 2003 for a fascinating study of fads in capuchin monkeys, showing intergroup differences in social gestures that must result from delay or blocks to knowledge transmission).

If the distribution of great ape habits results from ecology and does not attest cultural processes, does this mean that transmission of knowledge from one generation to another is unimportant? If most of the habits have been invented more than once, as their distribution implies, might they have been invented anew by each individual without social input? The likelihood of this uber-skeptical suggestion being true depends entirely on the *improbability* of invention. For clubbing with a stick, picking marrow out of a broken bone, or slapping a branch to attract attention, it is surely highly likely. However, some of the food-processing behaviors of chimpanzees, orangutans, and gorillas are remarkably elaborate. By "remarkable" we generally mean improbable of discovery, and to learn improbable things you need help. For an ape in the forest, the only help at hand comes by means of social learning. As the complexity of a skill increases, the likelihood of wholly individual learning decreases (Byrne

2007). Compare, for instance, the human traits of "liking pistachio nuts" and "raising water by making a shadoof." Individual exploration and trial-and-error learning are just not likely to be a complete explanation for an individual person's ability to make a shadoof, although of course both contribute. For this reason, it would not be sensible to require the same degree of evidence for social transmission of knowledge in shadoof making as for liking pistachio nuts, an easily learnt dietary choice.

Of course, great apes make nothing as complex as a shadoof. But part of the reason for interest in "ape culture" is the hope that it might help us understand the origins of human technological supremacy—which certainly does result from cultural accumulation of knowledge. It would therefore make sense to focus first on just those activities that manifestly require some skill. There is very much better evidence for social learning, of traits that show inter-population differences, in both rodents and fish than in chimpanzees or orangutans. Local differences in rat and fish behavior have been firmly established by experiments to be a result of knowledge transfer by social learning (Galef 1990, 2003; Helfman and Schultz 1984; Laland and Hoppit 2003; Warner 1988): rats dive for mussels in some rivers but not others, reef fish prefer some coral heads for their mating displays but not others, and so on. Yet this work has not led to comparable claims of "rat culture" or "fish culture." None of these undoubtedly socially transmitted habits is closely relevant to the cumulative culture of human technology. The elaborate processes used by apes to gain access to some foods at least might be.

As van Schaik et al. (2003) point out, the socially transmitted (the term they prefer is "cultural") contribution to an animal's behavior may contain a number of very different sorts of information. At the most basic, this may simply concern whether something is edible or where females tend to be when spawning: such things are socially acquired in a wide range of taxa. Song dialects in oscine passerine birds also give clear evidence of socially mediated patterning. These birds reliably acquire their species' song by social learning. In contrast to these widely distributed kinds of social transmission, van Schaik et al. assert that the social transmission of *skills* is unique to the great apes. If "ape culture" is worthy of special attention—beyond that accorded to social-transmission processes in more easily studied species such as birds and fish—then it is because great ape feeding ecology is reliant on skills sufficiently complex that they may need to be acquired socially (Byrne 2007). In general, local patchiness of distribution does not single out traits of manifest difficulty: few of the dozens of reportedly cultural traits of great apes involve any real skill, in terms of subtlety, complexity, and likely difficulty of acquisition, without a skillful model. Can we do better?

The skills of ape culture

If we were asked to look around for an example of a complex skill, we'd probably pick something that involved using tools or technology. An immense range of animal species has been recorded to use a tool; and a surprising number of species use tools regularly (Beck 1980; Shumaker et al. 2011). Elephants use sticks to scratch awkward places on their body, Galapagos woodpecker finches pull off cactus spines and hold them in their short bills to probe deeply into tree-bark for insects, hermit crabs live permanently in mollusk shells, and octopus may temporarily hide under them; the list goes on and on. But in each of these cases, a single type of tool is used for a particular purpose: each species is a "one-trick pony" when it comes to tool use. In many populations of chimpanzee, and one of orangutans, tool use goes well beyond the simple movements involved in those cases, because an intricate series of actions is involved in selecting, constructing, and employing the tools. Consider chimpanzee termite-fishing. Fishing is not a response triggered by the perception of edible termites: the insects are hidden deep within termite mounds, which have no visible entrances. The ape must know that at a certain season it becomes possible to pick open a hole in certain places on the mound. Opening a hole is in fact only possible above what will later become the nocturnal emergence tunnel for the sexual forms of the termites, so the chimpanzee must also know how to recognize a sealed exit. Only when a suitably long, thin, and flexible plant stem is inserted slowly through this entrance can it be withdrawn with termites attached. And plants don't just come that way; the stems have to be picked, stripped of leaves, and bitten to a standard length. Since the termite mound and suitable plant material may be separated by a considerable distance, tool preparation sometimes has to be done well in advance of arrival at the mound, and when it is not even in sight (Byrne et al. 2013; Goodall 1986). One population of orangutans has shown similar abilities, probing for honey with a tool held in the mouth, and eating *Neesia* fruits. *Neesia* seeds are embedded in a mass of irritating hairs. When orangutans eat *Neesia*, they first make a small tool by biting off a twig and stripping it of bark, then use it to scrape out the irritating hairs from a ripe, part-open *Neesia* fruit; when the hairs are cleaned away, the same tool is used to dislodge the seeds to eat (van Schaik et al. 1996).

Termite fishing and *Neesia* feeding share a characteristic that is just what we would expect from traditional transmission of technological information: *intricate complexity* of the behavior, because that's where social learning is needed. Intricately complex behavior patterns are highly unlikely to be invented multiple times, making cultural transmission an essential feature of their widespread dissemination. A similar case can be made in the orangutan for hole probing for honey; and in the chimpanzee, for hole-chiseling to access honey or ants, and for hammer-and-anvil use for nut cracking, and many other skills. They are the aspects that make great apes of special interest: ape culture is "special" because it allows transmission of behavior that is in principle relevant to understanding the origins of human culture, where it is the intricate patterns of complex action that have long impressed anthropologists.

If we focus on intricate complexity, instead of local variations without ecological correlate, some ape skills previously sidelined or rejected in the culture debate emerge as likely candidates for traditional transmission of technological information. As noted, ant dipping by chimpanzees varies in style in ways that make good ecological sense as ecological adaptations to different ant species (Humle and Matsuzawa 2002), but this need not disqualify the behavior itself as cultural. The elaborate series of skilled actions involved in ant dipping is highly unlikely to be learnt by a solitary chimpanzee. Of course, many of the important details of how the job is done are probably discovered individually, as with any complex skill: it is most likely only the "gist" of ant dipping with a stick that is socially learnt. The same may apply to several other tool-using skills of chimpanzees (e.g., nest construction, algae scooping, leaf sponging), not considered as part of chimpanzee culture, and to plant-preparation skills not involving tool using, such as opening and eating *Saba* fruit (Corp and Byrne 2002a). Almost all orangutan populations lack regular tool use, but most show intricate, complex skills for dealing with plants, in particular to enable pith consumption from spiny palms and rattans (Russon 1998). Gorillas do not make tools in the wild, but one population has been studied at the level of detail needed to detect skills of intricate complexity in plant processing. Several of their food-processing skills consist of highly structured, multi-stage sequences of bimanual action, hierarchically organized and flexibly adjusted to plants of highly specific local distribution (Byrne and Byrne 1993; Byrne et al. 2001a,b). In terms of the number and dexterity of operations that need to be performed, gorilla plant-processing actually exceeds anything yet described in chimpanzees. Gorillas, like chimpanzees and orangutans, apparently sometimes rely for their survival on elaborate, deft, and intricate feeding skills that are highly unlikely ever to be discovered by a solitary individual (Byrne 1997b, 2004).

The underlying cognitive ability for acquiring skills of intricate complexity, that these great ape genera—*Pan, Pongo,* and *Gorilla*—display in their different ways, is likely to have a relatively recent origin (i.e., in the ancestor of the living great apes), and relates directly to the origins of human technological superiority. Nothing quite like this has been reported in any monkey species, let alone any other animal species. But exactly what cognitive ability apes must have to copy these "improbable" skills, and whether this includes some insight, will be deferred to Chapter 11 where we will approach the issues from another direction.

For now, it must just be remarked that culture in human life includes other characteristics than the accumulation of knowledge beyond what any one individual could discover or invent. If this were all that human culture consisted of, then new ways of doing things would always spread unfettered round the globe, limited only by physical barriers. Where there were alternative ways of doing things the most efficient would eventually dominate. We'd all speak the same (best) language, we'd all eat particular foods in the same (best) way, given the same materials we'd all build houses the same (best) way . . . and so on. Manifestly, though, these things only happen to a limited extent.

Human cultural uniqueness

In humans, there are social barriers to the spread of knowledge and habits: the barriers often coincide with people's social identity as group members. Much of the subject of social psychology is concerned with what makes individuals identify with one group rather than another. In-group identification leads to negative feelings toward out-groups, and when those negative feelings become outright xenophobia they can lead to war and genocide. Most human cultural differences are best understood as the result of in-group membership. This results in a striking pattern, typical of human culture but never found in non-human great apes: where a whole suite of differences co-occur between one group of people and another. People who build pagodas are likely to use chopsticks to eat rice, people who build water turbines to grind barley are likely to show fraternal polyandry, and so on. In chimpanzees, there are no such "packages" of correlated behavioral traits, and the level of behavioral difference is often as high between nearby study sites as distant ones.

Many anthropologists would consider that group identity is what culture is all about: the spread of technological habits is merely a superficial marker of cultural identity, which is really a matter of a person's feelings and knowledge about the social group to which they belong. As far as we know at present, non-human apes do not feel group identification. Chimpanzees certainly carry out lethal attacks on neighboring groups (Goodall et al. 1979; Watts and Mitani 2001), and their behavior has impressed observers with its resemblance to our own: terms like "commando raids" and "warfare" are used. But there is no evidence that a chimpanzee who knows how to use a hammer and an anvil to break hard nuts, or who habitually reverses his termite-fishing probe when it is frayed rather than throwing it away, feels superior to those chimpanzees who can't break the nuts and who throw away their frayed probes.

Summary

Although the likelihood of innovation is higher in larger-brained species, animal innovation seems to be based on chance discoveries. Individuals of larger-brained species are more capable of representing the key variables of situations, and so more often profit from occasional lucky events by rapid learning. In theory, innovation is an alternative strategy to social learning, beneficial in quick-changing and unpredictable environments; in practice, the most innovative (and large-brained) species are often also those that rely the most on social learning. This includes the great apes, a group which has been often noted to be paradoxically conservative in adopting new habits. The conservatism of social learning results in animal traditions, such as the feeding habits of birds, where the spread of novel skills in response to new ways of artificial feeding can be observed in any suburban garden. The social learning that permits this kind of animal culture may be as simple as an automatic tendency to notice and represent what other animals in the social group are doing, with what, and where, thus priming the brain records of those places, things, and actions. The result is that individual exploration is focused on just those places, things, and actions that are most likely to be worth trying, so the habit spreads. As with innovation, no insight into mechanism seems necessary for these useful traits to operate.

Although the evidence that ape behavior relies on social learning is in general much weaker than for many other species, great apes are unique in that their repertoire includes learned skills for which individual discovery is highly improbable. Typically, these are feeding techniques which involve several stages, coordinated actions of the two hands, often involving tool use—and in general they strike observers as being surprisingly "clever" and "complex." The cognitive abilities that allow individuals to copy these improbable organizations of behavior will be picked up again in Chapter 9 and developed fully in Chapter 11.

Before we do, there is one very large body of evidence that needs to be considered: that from the research directly aimed at asking whether animals have insight into the social behavior of others; that is, do they understand the reasons why other individuals do what they do?

Chapter 8

Theory of mind

Understanding what others think about the world

"Theory of mind," or ToM for short, is the most familiar term in psychology for the ability to show insight into other people's minds; it was first introduced as a question about a human-reared chimpanzee: Does the chimpanzee have a theory of mind? (Premack and Woodruff 1978). By theory of mind, David Premack and Geoffrey Woodruff meant the ability to understand what someone else might know or not know. This is a remarkable capacity that most of us take for granted: we realize that everyone has had different opportunities to learn things, with different experiences, emotions, and reactions, so may be more or less knowledgeable than ourselves; and moreover, we understand that anyone's knowledge may be patchy or even completely incorrect. It enables us to reason about what they are likely to do, how they might be helped or hindered, and even to make moral judgments about things people do. None of the information we use for these decisions is "guaranteed reliable": our social knowledge is therefore a theory about someone else's mind.

Confusingly, there are several different but equivalent terms to describe this ability to show insight into social agency. The reason for this is that doing so is such an important ability that it has been discussed in several academic disciplines which don't always talk to each other, and each has coined its own term. Cognitive psychologists describe information in the mind as "representations" of reality, and since theory-of-mind knowledge is a representation of other representations (of the world), they use the term meta-representation as synonymous with theory of mind. The ability can even be applied to oneself: you not only have knowledge and hold beliefs, which vary in their reliability, but you can also estimate how likely that information is to be accurate. The general term for this in psychology is metacognition. Social psychologists are not concerned with whether or not people are able to represent the contents of other minds: since people normally can do so. In social psychology, the term for beliefs about others' minds is attribution, and we routinely "attribute" particular beliefs to others.

When distortions in beliefs about others' minds become systematic and prevalent, they may have profound social consequences, and attribution theory is about how that can happen. Philosophers have perhaps been concerned with this topic for longer than anyone else, and their term for it is intentionality. In philosophy, intentions include a lot more than the word's everyday sense, i.e., a goal to be achieved. Intention includes any mental correlate of an event or thing in the world outside the mind, real or imagined (events and things in the world are then called "extensions"). So, a specific memory about something that happened or might happen (episodic memory, technically speaking) is an intention. We can hold intentions about other's intentions; second-order intentionality is therefore equivalent to having a theory of mind. In the overlap of psychology and philosophy, the term second-order representation is used to include our beliefs about other minds and about our own.

Whatever you call it, the implications are the same: we can think about thoughts. Perhaps the clearest term is mentalizing, which has the advantage of stressing the process aspect of the ability rather than the mental states themselves. I shall employ it here as often as the more famous "ToM," a term which is hard to avoid altogether. I will examine a range of phenomena in which mentalizing is potentially involved in understanding social agents: their feelings as well as their knowledge, of the self as well as of others, in cooperation as well as competition.

Do any animals know about the thoughts of others?

Often treated as the cognitive Rubicon between humans and other animals, the ability to interpret other people's actions in an intentional way—as caused by their thoughts: their desires, knowledge, and beliefs—has been subject to immense research activity over the 30-odd years since Premack and Woodruff first asked whether the chimpanzee has a theory of mind (Call and Tomasello 2008). Throughout most of that period, there has been something of a gulf between the positive conclusions of those who observe natural animal behavior, and the doubts of those who carry out controlled laboratory studies (de Waal 1991).

Of course, over-interpreting the richness of naturalistic observation is not difficult! As humans, we are used to attributing complex thoughts on the basis of sparse clues given by thoughtful creatures, other people; and we may sometimes unthinkingly make the same attributions even to inanimate entities, like "malign" computer software. The famous *Fawlty Towers* scene in which Basil disciplines a misbehaving car comes to mind. Conversely, experiments may fail for many reasons other than the cognitive limitations of the subjects: but it is tempting to take negative evidence as evidence of inability. So some gap is to be expected—but in this case, the gap seems to have become particularly wide. For instance, analyses of primate behavioral deception, observed in the 1980s under natural conditions (Byrne and Whiten 1985, 1990; Whiten and Byrne 1988b), concluded that although most records could be explained as rapid learning in social contexts, some cases implied much more. Great apes in particular were considered to have represented the ignorance, knowledge, and false beliefs of other individuals (Byrne and Whiten 1992; for chimpanzees, specifically, see also de Waal 1982, 1986). These proposals ran completely counter to established views, and years later authoritative reviews concluded that even great apes were totally unable to represent any mental state of another individual (Heyes 1998; Povinelli et al. 1994; Povinelli et al. 2000; Tomasello and Call 1997). But the prolonged controversy also led to new empirical studies, which eventually brought observers and experimenters into register.

An early attempt to investigate whether animals could understand the knowledge of others was the "guesser–knower" experiment we met in Chapter 5, in which the subject is confronted with conflicting hints as to which of several places has been baited with food: from one individual who was visibly present during the baiting and able to see what went on—the knower; and by one who was not—the guesser. As we saw, this design suffered from several weaknesses: rewarded test trials could have allowed rule learning even during transfer tests; using humans as knowledge sources risked a bias in favor of closer relatives to humans; and the fact that the people knew the correct answer risked a Clever Hans artifact. The guesser–knower design depends on the subjects taking cues from helpful onlookers, even expecting help from others. In fact, chimpanzees and indeed most non-human primates may have little natural experience with such cooperative interactions (Hare and Tomasello 2004). When Hare and co-workers designed a competitive perspective-taking task of a more natural kind, quite different results were obtained. With this design, discussed in Chapter 5, chimpanzees clearly showed that, by witnessing the situation from their own perspective, they were able to compute what a competitor could see. At last, this was experimental evidence consistent with that from fieldwork: the understanding of geometric perspective taking. More excitingly, when the competitor was swapped, mid-experiment, for another, equally dominant individual, the subjects took account of the newcomer's lack of prior opportunity to see what had happened and consequent ignorance of it (Hare et al. 2001). Subjects were apparently able to represent what others had previously seen—in other words, what they knew, as well as what they could currently see. Although some still dispute this interpretation (Karin-D'Arcy and Povinelli 2002; Povinelli and Vonk 2003), considerable converging evidence now supports the contention that chimpanzees can appreciate the knowledge or the ignorance of other individuals (Call and Tomasello 2008; Tomasello et al. 2003).

When tested in a similar competitive situation, rhesus monkeys also showed understanding of the geometric perspective of others: if they could steal a grape from one of two humans, they reliably chose the person who was unable to see their action (Flombaum and Santos 2005). Both monkeys and chimpanzees also proved able to compute the consequences of hearing, preferring to take food from "quiet" containers that will not alert an inattentive competitor to their theft (Melis et al. 2006a; Santos et al. 2006). Perhaps most surprisingly of all, some corvine birds also proved able to deduce and remember whether a competitor is ignorant or knowledgeable, on the basis of what they have been able to see: we have already seen this evidence, in Chapter 5, in the behavior of the western scrub jay and the northern raven. Both species show an elaborate range of tactics that appear designed to avoid pilfering of their caches, and deploy them specifically against individuals who have been in a position to see them making the caches (see reviews in Bugnyar 2007; Dally et al. 2010; Emery and Clayton 2004). Tactics include waiting until potential pilferers are distracted before caching, making false caches, leading competitors away, re-caching food when alone, and hiding food behind barriers or in poorly lit locations. Critically, scrub jays who are naïve to pilfering do not show re-caching of food after their caching has been observed, but once they have been pilferers themselves, they do (Emery and Clayton 2004). All these data are entirely consistent with an understanding that others' behavior is driven by their knowledge, and that their knowledge depends on what they have been able to perceive.

Although many non-human species may therefore be sensitive to the potential knowledge or ignorance of their competitors, it is much less clear whether any of them understand the concept of false belief. Part of the problem is that experiments with non-verbal subjects can easily become very complex, which itself may make for failure. Kaminski and colleagues attempted to tackle the question of false belief by using a competitive game between two individuals, either children or chimpanzees, in which the subject could make only one choice in each trial. They could either opt for turning over one of three cups, having seen a high-quality food item placed under one of them, or instead choose a low-quality reward that was always available (Kaminski et al. 2008). Getting only a low-quality reward is clearly a poor option, but it might be a safe "cop out" if the task gets too difficult! The two subjects took it in turn to choose first, and the second could not see what the first had chosen: so the high-quality item might already have been taken by then. When they had seen a cup baited with a high-quality food, and also seen that the other individual saw the same baiting process, both 6-year-old children and chimpanzees wisely avoided that choice when it was their turn. Both species therefore showed differentiation between knowledge and ignorance. In a second experiment, the experimenter picked up the food after the baiting process, and either put it back or moved it to a new location: conditions differed in whether both subjects, or only the one who chose second, saw this extra move. If you choose second, the smart strategy is to go for the high-quality food if you alone saw it put back, since the first subject should have falsely believed that the food was elsewhere—and so wasted their turn. Six-year-old children used this strategy, showing that the task was not too complicated for them to deploy their understanding of false beliefs in its solution. The chimpanzees did not; nor did 3-year-old children. However, other work suggests that babies much younger than 3 years can represent false beliefs (Onishi and Baillargeon 2005; Southgate et al. 2007), so the failure of the 3-year-old children—and thus also of the chimpanzees—in Kaminski et al.'s task may reflect confusion from the task's complexity.

Similar negative findings were obtained with free-ranging rhesus monkeys, using a simpler experiment (Ruiz et al. 2010; Santos et al. 2007). Santos and colleagues used an expectancy-violation design, treating looking time as an indication of surprise, based on a task originally devised for human infants (Onishi and Baillargeon 2005). Free-ranging monkey subjects were presented with a table on which a plastic lemon was able to move on a track, from side to side. In the critical trials, the human presenter was unable to see some of the movements of the lemon because of an occluder, although the monkey could. Thus, if the occluder was absent and the (human) presenter watched closely while the lemon moved to a new location, they should have a true belief of its position; but if the occluder blocked their view during the same move, they should have a false belief (out-of-date information) about the current position. When monkeys watched the presenter's subsequent search for the lemon, their looking times showed they expected the presenter to search in the right location in the true belief condition, and were surprised (longer looking time) if they searched elsewhere. However, in the false belief condition, looking times did not differ: monkeys, unlike human infants, appeared not to have any expectations in this case (Santos et al. 2007). Perhaps the only animal able to compute the consequences of a false belief in another is the human.

If so, a conflict remains between observational and experimental evidence, albeit a much more limited one. The analysis of deceptive behavior in primates suggested that the great apes, although not monkeys or prosimians, were potentially able to represent false beliefs (Byrne and Whiten 1992). In a number of different ways, particular ape individuals were recorded deploying tactics that functioned only when the target competitor was led to hold a false belief. Unlike most deceptive tactics of primates, these were simply not plausibly explained "away" as the product of learning from repeated past experiences (Byrne 1993; Byrne and Whiten 1991), because, for instance, the behavior deployed was something never noted in the species in other contexts, or the only circumstances that might have allowed learning were highly dangerous ones. Resolution is clearly needed: either, despite the apparent implausibility, these tactics were learnt without any understanding of false belief; or, despite the efforts of researchers to design a readily comprehensible experiment, the sheer complexity of those used so far with great apes may have prevented the subjects showing their real abilities. At present, there is positive evidence that chimpanzees and rhesus monkeys can show insight into another's ignorance or knowledge; but the possibility that any non-human can compute another's false beliefs has not been confirmed, despite several experiments designed to do so.

Understanding another's role in cooperation

When we plan to cooperate, it is critical to understand the contribution others can make to the joint activity, and this kind of understanding relies on mentalizing about other people's knowledge, abilities, and likelihood of helping. Cooperation can also happen almost accidentally, without those levels of social understanding. When several chimpanzees block all the escape routes from a tree, and thereby capture and kill an arboreal monkey, there is no doubt that the hunt feels like a cooperative venture to the monkey. But the chimpanzees may not represent the situation that way: perhaps each individual hunter is simply going to the best-bet escape route that the monkey could take, which maximizes its own selfish chances of capture? This hypothesis was put forward over thirty years ago (Busse 1976), but even today researchers are divided as to whether there is more cooperation than this in chimpanzee hunting. Some describe individuals taking different roles—driver, blocker, ambusher—in a richly cooperative team effort; others conclude that in dense tropical forest and the confusion of a hunt, getting real evidence of cooperative hunting is just not possible, even if it does happen (Boesch and Boesch-Achermann 2000; Mitani and Watts 2001).

Captivity presents much better opportunities for detecting whether animals understand cooperation. As long ago as 1937, Meredith Crawford developed a simple two-role cooperation task for chimpanzees (Crawford 1937), and Daniel Povinelli has revisited this classic study (Povinelli and O'Neill 2000). A box that is too heavy to be pulled by one individual has two ropes attached, and both can be reached, but not at the same time by a single chimpanzee. What is needed is for two chimpanzees to grab one rope each and pull together to get the food that is placed on the box. Because the ropes are threaded through hoops, if one individual pulls prematurely the rope simply slips through and detaches from the box, so it is crucial to wait until the other individual is ready and willing to pull. Although chimpanzees did learn to perform the task, levels of effective collaboration were modest when individuals were not explicitly trained how to do it. Moreover, even trained "expert" chimpanzees did not spontaneously gesture toward naïve individuals to direct their attention to the key features of the task; and while they sometimes waited for the naïve chimpanzees to approach the ropes, they then grabbed and pulled in their usual way rather than synchronizing with the other. But negative data are always ambiguous: maybe there was some motivational reason for the chimpanzees' unimpressive level of performance? A more recent study found that chimpanzees not only recruited a partner when they needed one for the rope-pulling task, but also reliably chose the better collaborator (Melis et al. 2006b). If no helper were needed, chimpanzees did not unlock the door for a potential partner (and food competitor!): they only did so when two collaborators were essential. When they'd had no experience of a collaborator's effectiveness, they had no preference for whom they released to help them; but once they'd seen some good and bad collaborators, they homed in on the effective ones to release and left others locked in. Asian elephants, tested on a scaled-up version of this task, readily learned to wait for their partner to arrive at a place where it could grasp the rope with its trunk before pulling themselves (Plotnik et al. 2011); whether they also distinguish among potential collaborators according to their reliability has yet to be tested. In other circumstances, chimpanzees often act as if they understand a human's need for them to help: for instance, when an experimenter apparently dropped an object and struggled to reach it, a nearby chimpanzee would often pick it up and hand it over (Warneken and Tomasello 2006; Warneken et al. 2007). But this evidence may be less compelling than it seems, because in most zoos or research stations, apes are regularly rewarded for picking up and handing over objects that keepers have accidentally dropped into the enclosures, so the subjects may have had previous (rewarded) practice.

Finally, chimpanzees show some understanding of the nature of collaboration itself. In a classic two-role challenge, devised by William Mason, one individual can see which of several pairs of containers has been baited with food, but cannot do anything about it. The other participant has access to a series of handles, and if he pulls the right handle both participants will get access to a baited container, but he cannot see which ones are baited. Over repeated trials, monkeys learn to signal which pair of containers to go for, if they can see; or learn to pull the handle indicated by the other monkey, if they cannot (Mason and Hollis 1962). Daniel Povinelli modified the task to find out whether each participant had understood the cooperative nature of the task, or simply learned an approach that worked. After having learnt one of the roles, if they are switched to the other role monkeys had to start again and learn the new role, as if from scratch (Povinelli et al. 1992b). Chimpanzees, like monkeys, could learn either role of the task, but crucially if their role was then switched, they hit the ground running and needed no extra trials to get it right (Povinelli et al. 1992a). It seems that whichever role a chimpanzee happens to learn to do as an active participant, what they learn is the logic of the task: how the cooperation works, presumably including the knowledge that only one of the two chimpanzees is in a position to see—and therefore know—which handle would be best to pull.

Understanding oneself

When confronted with a mirror for the first time, most animals treat the image as an unfamiliar conspecific: they make inappropriate "social" responses and often try to find where "the other animal" is, by looking behind the glass. Domestic cats show these reactions, young children show them, and adult humans who have never in their lives had experience of mirrors also show them—as was vividly demonstrated during some of the "first contacts" between New Guinea highland people and Western explorers during the twentieth century. With experience, however, reactions change: all humans come to realize that the image is of themselves, a result of reflection. At this point they begin to use the mirror in distinctive new ways: examining parts of their face that they cannot see directly, or using the mirror as a periscope or to see around corners where their head cannot reach. Cats, on the other hand, do not: their social responses may habituate, but then they simply ignore the mirror, and show no sign of understanding the reflected images.

Nor are cats alone in this failure of understanding. In fact, even given extensive experience in which to discover the mirror's properties, almost all animals fail to make mirror-appropriate responses. Some birds continue to make social responses, day after day, trying to "drive off" the intruder in their territory. Most animals, like the cats, just avoid looking at the mirror after a series of attempts to make social contact with the "animal in the mirror." There was therefore great excitement when it was found that a few species behave more like humans, apparently understanding the image as of themselves. But why is this so hard, anyway? Especially for an animal with paws (or hands) that they can see in action, there is continual opportunity to see that the paw in the mirror appears the same as their own, its movements perfectly match their own, and when their paw touches the glass so does the paw in the mirror. How hard can it be to associate the two? And if paws (or hands) are realized to be in certain ways identical to their images, would that not be expected to generalize to the rest of the body with which the paws are connected? On the face of it, the task should be easy: why isn't it?

For most species, their behavior gives few clues to help us. But in the case of monkeys, we can rule out one possibility: that the mechanics of reflection pose the insuperable problem. Monkeys can learn to understand reflection, provided it is the reflection of something other than themselves. If they catch sight in the mirror of a dominant conspecific, they begin to make submissive sounds and facial displays before turning round; and if they are set a task that depends on reaching accurately to places they cannot see directly, they soon learn to benefit from a reflected image in a mirror (Anderson 1984). With more practice, monkeys can adapt to using real-time video images (i.e., not L–R reversed), upside-down images, or even a medley in which every trial is different: reversed or not, upside-down or veridical (McKiggan 1995). With a quick wave of their hand in front of the camera, they judge which transposition they are presented with, and use it efficiently. Yet no monkey has ever convincingly been shown to react appropriately to their own face in the mirror. It may be that many species cannot understand the everyday physics of mirrors, but monkeys evidently can and it still doesn't remove the barrier to understanding their own faces.

Whatever that barrier is, surmounting it is not (quite) unique to humans. Gordon Gallup noticed that chimpanzees, after they became well used to the presence of good-quality mirrors in their enclosure over the course of weeks, began to make different responses than their original social ones. They would make rhythmic movements of arm or body, watching their image intently, repeatedly touch the mirror, and sometimes appear to contort their faces or open their mouths as if to see areas that they could never view without a mirror. It seemed that their experience had led to a human-like understanding of their appearance, and to check, he devised an experimental method: the mark test (Gallup 1970). In this, a subject is surreptitiously marked (for instance, when anesthetized for a veterinary procedure; or, if the animal is used to being touched, by wiping a cloth over the face with actors' make-up concealed in it). Marks are placed in one obvious place and one place visible only by way of a mirror. No mirror is allowed in the animal's view for the first 30 minutes, during which time it should find and examine the obvious mark—and crucially not find or examine the out-of-sight one. Then a mirror is replaced: if the animal has really understood that the image is of itself, once it catches sight of its reflection in the mirror it should immediately reach to and investigate the hidden mark by hand. Many chimpanzees, and a similar proportion of bonobos, gorillas, and orangutans, do just that: much of this material is available on video, and vividly depicts the fact that great apes can readily learn to understand the action of mirror reflection on their own bodies just as we do. Magpies (*Pica pica*), a corvid species, have been tested in just the same way as these apes, and showed very similar patterns (Prior et al. 2008). All birds originally gave social responses to mirrors, but in some individuals these were replaced by repeated movements of the body while in front of the mirror, as if testing contingencies; and when tested with the mark test these individuals showed mark-directed behavior specifically when in front of the mirror. Bottlenose dolphins, lacking anything equivalent to the hand, nevertheless show strong signs of understanding their mirror reflections, by twisting their necks and positioning their bodies in the water in ways that only make sense if they are using the mirror to examine their hidden parts, just as mirror-experienced apes do (Reiss and Marino 2001). The only other animal for which mirror self-recognition has been reported is the Asian elephant (*Elephas maximus*). Although there have also been several negative reports, one zoo-housed elephant did show distinctly different self-directed behaviors in the presence of a mirror, specifically after surreptitious marking (Plotnik et al. 2006). These reactions were not immediate but began several minutes after the elephant saw itself in a mirror, and were first given when actually away from the mirror—very different characteristics than those shown by chimpanzees. However, when a larger sample of elephants was later tested, at least 1 and possibly 3 out of 22 showed behavior exactly like that of mirror-sophisticated chimpanzees (Josh Plotnik, personal communication). Regardless of which animals eventually prove to be able to understand that a face reflected in a mirror is their own, the fact remains: what is really strange is that most other animals fail in a task that seems so simple to learn.

The explanation originally suggested by Gallup (1979) is controversial, yet I believe it to be the correct one. He suggested that in order to be in a position to realize that the animal in the mirror was itself, the animal must first have a *sense* of self: in cognitive terms, we would say a mental representation of the self. What the individuals of most species lack, then, are any representations of themselves— as others would see them. Without that, Gallup went on, these animals could have no conception of what might one day befall them (death), of their own mortality, even though they would have ample opportunity under natural conditions to observe death in others. This provocative speculation has been ignored by most researchers, presumably because of the obvious difficulty of testing it, and strongly criticized by some (e.g., Heyes 1994, 1995). Yet a subsequent theory has provided a way in which just such an ability—to see the self as others see us— might have developed, specifically in the ancestor of the living great apes but not in other primates or most other species (Povinelli and Cant 1995).

The starting point for this theory is the belief that the common ancestor of living great apes was quite like an orangutan, which has some support in the fossil record (Pilbeam and Smith 1981). If so, it might well have shared a major problem with today's orangutans: the difficulty of arboreal clambering for a heavy animal. Small, light quadrupeds like monkeys run on top of branches; when they occasionally have to leap from one tree to another, they can be pretty confident of grabbing a branch capable of bearing their weight. Gibbons travel entirely by branch swinging, with consummate skill, but they still make mistakes, and 30 percent of recovered gibbon skeletons have healed breaks. Orangutans are far too big to make mistakes like that and survive, so they clamber sedately, "four-handed," through the tangled vegetation of lianas and branches. However, the weight of the orangutan also causes the tangled vegetation to deform as they move, so precisely which routes are available to them is not obvious—until they try, and then it may be too late. Thus, according to the locomotion expert John Cant, the orangutan must be able to simulate the effects that different (future) travel decisions will have, without committing to them: an engineering problem, in which the body must be factored into a complex network of relationships. Cant collaborated with Daniel Povinelli, a cognitive researcher puzzled by the distribution of mirror understanding, and they realized that this might explain the puzzle (Povinelli and Cant 1995). The ability to "de-center," and so to represent the self as an entity (which is just as others see it), is essential for orangutan survival. So, a selection pressure for safe arboreal locomotion in an orangutan-like, ancestral great ape could have had the "spin off" of conferring the ability to represent the self in other contexts, irrelevant in the wild, such as when reflected in mirrors.

This theory makes a clear prediction about which species should show mirror self-recognition—only those for which treating their own body in movement as a complex engineering problem is critical for survival—so the distribution of this esoteric ability becomes important for testing it, and maybe understanding how we first acquired a sense of ourselves. With hindsight, a partial case can be made that the sparse data on mirror understanding is consistent with the theory in some species: in toothed whales (their movements and coordination with others have caused admiration and amazement so often, and accurate planned movement must surely be critical for their survival), and elephants (whose vast bulk does not deter them from rapid movement and travel on steep slopes and rocky mountains); but not, surely, in magpies. If the theory proves to be incorrect, then at present there is no other ready to replace it!

Gallup also asserted that mirror self-recognition is an index of the kind of self-understanding that allows one to comprehend mortality: and perhaps this may be tested after all. Is mirror understanding by animals associated with any signs of understanding what happens when an animal dies?

Understanding death

Death is pervasive in nature: all individuals die, and many cause the death of others for their own survival. None of this requires that any individual should have insight into what it means to die. Provided they are equipped with fundamental traits that help them avoid dying, and learn well any aspects of the environment that will help their survival, no more is needed. Worrying about death is not helpful for living well, as many people have remarked over the ages. It has therefore never surprised biologists if individuals of their study species appear not to do so.

To take a concrete example, consider red deer. A local Fife deer farm, concerned at the quality of captive animal welfare they can provide, decided to slaughter the deer on site, avoiding the stress of travel to a distant licensed abattoir. The deer are fed daily, from the back of a tractor and trailer: when one is to be slaughtered, it is shot through the head with a rifle from the tractor dispensing food. The others in the group, who have been together all their lives and may include close relatives, show almost no reaction (John Fletcher, personal communication). Some deer may startle slightly at the bang (and when young they no doubt took time to get used to the loud noise of a gun at close quarters), but then they go on feeding peacefully, following the tractor that is dispensing forage as the body of their companion is gradually left behind on the ground behind them and retrieved for butchery. Red deer, like many animals, are exquisitely sensitive to cues of danger, and react dramatically to signals of fear or distress from others, but there is none from a deer shot through the brain. Deer do not react to death: it appears not to be something they think about at all.

When animals react in more "human-like" or "unusual" ways to the death of conspecifics, then, it causes comment. Probably the most frequent case of an unusual reaction to death is the treatment of dead babies by their mothers. Stillborn primate infants are often carried or groomed by their mothers, in monkeys and especially great apes, and this behavior is also sometimes seen in carnivores like leopard (*Panthera pardus*); but it is difficult to interpret these observations. Superficially, indeed, they seem to suggest a dramatic failure to understand what has happened. That was certainly my own impression when I had the opportunity to witness the treatment of a dead newborn (or stillborn) baby by a young gorilla mother. On the first day, the mother cradled the infant as if it were alive, and her relatives and unrelated youngsters clustered around her to look: it was hard to work out that the infant was dead rather than dozing, and an error on the mother's part seemed reasonable. As hours and days went by, however, interest by other individuals waned; indeed, the mother seemed to be shunned, perhaps because the corpse smelled badly, although she still trailed it about by one limb, banging on the ground behind her. After a few days, she abandoned it and went back to playing in a relaxed way with other young gorillas in the group, as if she had no care in the world. It seemed to me that this showed a very un-human failure to understand death; but I was soon disillusioned when I found out about the equivalent human case. The reality is that the most effective medical approach to the trauma of a stillbirth in humans, discovered thanks to pioneering work in Australia, is to allow the mother to keep, cuddle, name and show off her baby over a period of days. Mothers with a rare genetic condition that causes a high probability of stillbirth, but no deficit to any live births, have had to endure this trauma repeatedly. They report that the new, "gorilla-like" method allows them to adjust and mourn, and then put away from their minds the lost child; the original, "hygienic" approach of never seeing the body left them with nightmares for life. Do non-human apes and monkeys mourn and adjust by their natural carrying-behavior? It is very hard to find out for sure.

The behavior of elephants has always stood out in regard to death, because they show a special interest in the bodies of their kind, as well as showing empathic and helpful reactions toward distressed or dying individuals (Bates et al. 2008; Douglas-Hamilton et al. 2006); these behaviors are directed toward both kin and non-kin. They may use tusks, trunk, or feet in attempts to lift or carry sick, dying, or dead elephants, and males may attempt to mount a dead one; individuals have been seen to feed sick or even dead elephants, and are often noted to collect vegetation and soil and use it to cover an elephant corpse; and they may guard the body, keeping predators or other elephants away (Poole and Granli 2011). Stranger still, when the long-decayed remains of elephants are encountered, elephants typically investigate, explore, carry about, and play with the bones—or just quietly consider them. Whether these reactions are specific to remains of elephant has been investigated experimentally. Karen McComb and colleagues set out bones and other remains of elephants and bones of other large mammals, like hippopotamus and rhinoceros, so that groups of elephants would be likely to come across them; all the material was washed with detergent between each trial, to remove the chance that reactions were to the scent traces of other elephants. Stronger reactions of curiosity and exploration were shown to elephant bones and tusks compared with similar-sized bones of other species (McComb et al. 2006). However, since the detergent would have removed scent cues, it was not possible to know if reactions to natural remains would be different with known (dead) companions. While the elephant graveyard is a hoary myth, elephants do seem to show strong and unusual reactions to the death of a conspecific compared to most species.

A recent description of the death of a chimpanzee within a cohesive and long-established social group strongly suggests that this species, too, has some understanding of death (Anderson et al. 2010). The researchers described pre-death care, inspection of the body for signs of life, nighttime attendance at the corpse by relatives, and later avoidance of the site of death—patterns uncannily similar to our own normal reactions to the corpse of a dead friend or relative. (The reason that such behavior has only so recently been first described may be that zoos normally separate dying individuals from their social groups, in a belief that this will reduce their stress. The new data suggests that this may be exactly the wrong thing to do, and enlightened zoos may now change their policies.)

Just as with mirror self-recognition, the real mystery is why such reactions are so rare. Individuals of almost every social species will have opportunities to observe conspecific death and its consequences: why should they not understand the process? As Gallup suggested, just as it may be impossible to understand the face staring back from the mirror without a mental representation of the self, the possibility is that—without a representation of the self as an independent entity—the death of others has no personal significance, unless one is immediately dependent on their aid. If so, then those species likely to react with understanding to the death of others should be specifically those showing self-recognition. This hypothesis could be tested by further research on reactions to conspecific death, concentrating on species well known to be able or unable to recognize themselves in a mirror.

Empathy

Reactions to the death of companions and recognition of the self are phenomena that may also relate to a general capacity for empathy. Empathy is defined as the ability to share someone else's feelings or experiences by imagining what it would be like to be in their situation (*Cambridge English Dictionary*), often referred to as "putting oneself into another's shoes." Empathy is considered one aspect of human consciousness (Thompson 2001), and the ability to detect and respond appropriately to the emotions of others is a cornerstone of normal social function. The recent discovery of a mirror system for emotional responses in humans has provided evidence for the neurological basis of empathy (Jabbi et al. 2007; Keysers 2011; Wicker et al. 2003), but little is known about the evolution of this emotional mirror system and to what degree it is shared by any other species. Macaque monkeys are known to possess mirror neurons that react to the physical actions of others when they match actions in the monkeys' own repertoire (Gallese et al. 1996; Rizzolatti et al. 1996), but the analogous emotional mirror system has not yet been identified in non-human animals.

Simple forms of empathy, such as emotional contagion, have been used to explain contagious yawning, scratching, and the behavioral copying shown in the play and aggression of chimpanzees and Japanese macaques (Anderson et al. 2004; de Waal 2008; Parr et al. 2005). Lisa Parr and colleagues argue that "this type of emotional awareness functions to coordinate activity among group members, facilitate social cohesion and motivate conciliatory tendencies, and is likely to play a key role in coordinating social behaviors in large-brained social primates." That may be so, but behavioral contagion is also evident in chickens and all these phenomena can be explained with the simple concept of response facilitation (Byrne 1994; Hoppitt et al. 2007), which we first met in Chapter 7. Nevertheless, there is evidence that even behavior as seemingly simple as contagious yawning does correlate with empathic understanding in humans (Lehmann 1979), so behavioral contagion may well be a precursor to, or simplified form of, sophisticated empathic abilities.

Frans de Waal has reviewed chimpanzee behavior that gives evidence of higher levels of empathy (de Waal 2008; de Waal and Aureli 1996). First, a chimpanzee may show an affective response to the plight of another, showing signs consistent with feeling some of the emotion of the other: "sympathetic concern." Those who work closely with captive chimpanzees have often described their animals as behaving with apparent concern, going as far as a case of a male chimpanzee who lost his life in an effort to save a drowning infant; but the most regularly observed evidence of sympathetic concern is in "consolation" behavior (de Waal and van Roosmalen 1979). After two chimpanzees have been involved in aggressive conflict, an uninvolved bystander often goes over and embraces the loser, as if to reassure them after their unpleasant experience. Consolation has been reported in gorillas, bonobos, and young children, as well as chimpanzees, and also occurs in at least one corvid, the rook (*Corvus frugilegus*), where a bird's long-term mate will console them after they have been involved in aggression with a neighbor (Seed et al. 2007). However, consolation appears to be rare in monkeys, only reported so far in one species of leaf monkey (Arnold and Barton 2001).

Sometimes, empathy can involve taking the emotional perspective of another, understanding their needs and problems, and attributing mental states to them: a qualitatively higher level of cognitive computation. "Empathic perspective taking" might be best detected in animal behavior by the deployment of targeted help to another, in which the help was attuned to the other's problem but irrelevant to the helper. For instance, mother orangutans regularly use tree swinging to allow their infants to bridge arboreal gaps, for which they themselves have no need of such help. De Waal (2008) gives several examples of helping that showed awareness of another's limitations and needs from captive chimpanzees, including an adult who deliberately attracted the attention of keepers to the plight of several youngsters who were exploring a dry moat when water began to refill it. Similar reactions have been shown by silverback gorillas in several zoos, when young children fell into their enclosures and hurt themselves. To the amazement of the world's press (though not of most primate researchers), the gorillas protected the hurt children and carried them to where they could be retrieved by medical staff.

Empathy matching the levels shown in these great apes has also been described in a retrospective analysis of 35 years of observations of the Amboseli population of African elephants (*Loxodonta africana*) (Bates et al. 2008), which noted that the simpler forms of empathy, such as comforting and seeking physical contact with calves, can be observed every few minutes in Amboseli. Protection and assisting with mobility also occur on a daily basis. Sympathetic concern was shown by numerous instances in which elephants offer protection and comfort to the calves of others, babysit them, or retrieve them from harm. Empathic perspective taking, when appropriate help is targeted toward needy individuals, was shown in wild elephants by several cases in which calves were helped to overcome mobility problems, when the helper herself had no such problem. For instance, where a calf could not clamber out of a muddy waterhole, an adult, noticing its plight, neatly inserted one of her tusks in the ground in just such a way that the infant was able to hook its elbow round the tusk and pull itself out.

It would be convenient, for the theory that understanding of the self and empathy for others are related abilities, to report that there was excellent evidence of empathy in dolphins, also. Certainly there are many reports of behavior consistent with sympathetic concern for others and empathic perspective taking, but the practical difficulties of studying cetaceans under natural conditions mean that none is as compelling as those in elephants and apes. Some delphinids, like pilot whales (*Globicephala* sp.) and false killer whales (*Pseudorca* sp.), are prone to accidental stranding, and in these cases swimming whales often remain close to their stranded companions, despite the risk of remaining in shallow water. It is not easy to determine whether they are trying to reassure and perhaps help the stranded whales, or simply responding mechanically to distress calls, although there is some evidence that responses to distress calls are tuned to individual identities (see review by Kuczaj et al. 2001). Dolphins, since antiquity, have been known occasionally to rescue drowning seamen after boat disasters; the skeptic would point out that we are less likely to hear of any cases where dolphins have intervened to drown struggling swimmers! Whales of various species have been recorded keeping harpooned companions from drowning by supporting them at the surface or interposing themselves between the hunter and an injured conspecific (Caldwell and Caldwell 1966), but whether these behaviors are driven by empathic understanding is unknown. Intriguingly, the closest dry-land relative of the whales, the hippopotamus (*Hippopotamus amphibius*), has also been recorded behaving in an apparently empathic way. In two cases, a hippopotamus intervened to protect a small antelope: one attacked a crocodile that held an impala in its jaws, and when the impala was thus released it guarded it and licked its wounds; another pushed a tired swimming impala to shore when it had risked drowning (Leland 1997).

In sum, there is some reason to think that there may be a common basis to: (1) the cognitive capacities that allow certain species to recognize a mirror reflection as being of themselves; (2) the capacity underlying strange reactions to conspecific death that suggest some understanding of what death involves; and (3) the capacity to feel sympathy and recognize empathically the problems faced by others. Evidence is patchy, but the constellation of great apes, corvids, elephants, and cetaceans recurs in these areas; and the absence of data suggesting any such abilities in the vast majority of other animal species needs explanation.

Teaching

Education is so important to us—it is evidently so valuable to success in the human world and so much a part of constructing that world—that it is no surprise that there has long been fascination with the possibility that other species might also be able to teach. Anecdotes abound, but Tim Caro and Marc Hauser brought system to the collection of data by putting forward an operational definition of teaching that could be applied to animals (Caro and Hauser 1992). They specified that, for teaching to be proven to occur, an experienced individual must change its behavior in the presence of a naïve companion, in a way that provides no immediate benefit to either party, or even imposes short-term costs on itself, but causes the companion to learn. The thinking behind this definition is as follows. Naïve individuals can learn socially in many other ways: the key difference with teaching is that it is the experienced individual that changes its behavior to suit. If that supposed teacher benefited anyway, we could never be sure whether the naïve individual's learning was more than a side effect. If the naïve individual benefited directly in material ways, then we'd need no other explanation: even if learning did occur, it might have been incidental. And finally, the naïve animal must really learn something as a specific result of the teaching.

Caro and Hauser were aware that it was tough to satisfy all of these prescriptions; and in their paper they discuss a number of fascinating behaviors that "nearly" meet the definition, but not quite. Most strikingly, Caro's own work on cheetah (*Acinonyx jubatus*) and domestic cat prey-catching highlighted a number of characteristics that strongly suggest teaching (Caro 1980, 1994). Like domestic cats, mother cheetahs bring back disabled prey for their young to "play" with. In domestic cats, the experience gives the kittens a measurable advantage in the development of their hunting skills. In cheetahs, the degree of disablement is tuned to the age of the cubs: for the youngest litters, the prey is generally too injured to escape even the most incompetent approaches, but for older cubs the mother brings back lightly injured prey of species that easily escape from poor handling. Not killing the prey risks that the mother's hunting will be wasted and the young will go hungry, and confers no conceivable benefit to her or her cubs, unless it serves to teach. It is presumed that the young learn to capture prey from these experiences, but since no experiments could be done, some doubt remains.

Since then, experimental work on several species of very different kinds has discovered a few cases of teaching that meet every one of the operational criteria. Ants teach others the correct way to a site of exploitable food (Franks and Richardson 2006): ants are held and guided along the route by companions, if they show signs of not knowing the way, and helping in this way slows up the travel of the knowledgeable ants. Meerkats (*Suricata suricatta*) help their young to learn prey-handling skills by providing live prey, and they change what they offer in response to changes in the pups' begging-calls; this improves the pups' skills (Thornton and McAuliffe 2006). Pied babblers (*Turdoides bicolor*) teach their young to associate specific calls with food, by conditioning the young with food delivery as a reward (Raihani and Ridley 2008). The success of these experimental studies, in finally meeting the tough criteria set by Caro and Hauser, has led some researchers to vaunt "casting aside anthropocentric requirements for cognitive mechanisms assumed to underpin teaching in our own species" (Thornton and Raihani 2010), and to point out that teaching seems to occur widely across the animal kingdom, its occurrence driven by ecological need rather than cognitive capacity. If one's main interest is in the flow of information within social groups and how that relates to ecology, ignoring cognitive mechanism is fair enough; but much of the original excitement at the possibility that animals might teach was that it could give clues to the evolutionary origin of such an important human capacity. Avoiding anthropocentrism may therefore risk throwing the baby out with the bathwater.

None of the researchers on babblers, meerkats, or ants makes any claim that their subjects teach by means of insight into the lack of knowledge of their companions. Caro had explicitly pointed out that although his cheetahs modulated their prey injuring according to cub age, he could find no evidence that the mother assessed cub competence. Likewise, Thornton's meerkats changed their provisioning in response to the pup's calling, not their own assessments of individual pup ability. Pied babblers taught all their young, wholesale. And while the ants did assess competence in a sense, this was only a matter of physically taking the right route. Caro and Hauser's operational definition was deliberately based on functional, not intentional teaching—in order not to set the bar impossibly high, for picking candidate cases of teaching—but the real news would be intentional teaching in animals. To discover evidence of that, we may need to temporarily lower the bar a little (knowing, as we now do, that functional teaching is quite widespread in nature), in order to catch a wider range of cases among which some may prove to be based on insight into others' knowledge and competence (Byrne and Rapaport 2011).

Intentional teaching implies assessment of what is (not) in the current competence of the learner. The key question here is whether assessment is done at a superficial, perceptual level (as in the case of ant teaching), or by understanding the learner's knowledge and skill set. Focusing on cases where there is evidence that the teaching is done intentionally filters the data in a different way, and switches the focus to quite a different cast of actors. As might be expected, chimpanzees have several times been argued to teach intentionally, although the evidence is not strong. Of the nine long-term study sites where close-range data on chimpanzees have been collected for many years (for example, over fifty years at Gombe, Tanzania), teaching has been reported at only one. At the Taï forest, Ivory Coast, an 11-year study of nut cracking gave some evidence of teaching. Taï chimpanzees do not find nut cracking easy to acquire; it takes young animals years to learn how to crack the very hard *Panda* nuts, by bringing a rounded cobble down firmly on a nut positioned on a flat anvil rock—but avoiding smashing all the nut meat completely! Christophe Boesch described a number of ways in which mothers regularly facilitate the learning of their offspring (Boesch 1991). Mothers leave uncracked nuts by the anvil, which non-mothers never do; they allow the infant to use their own hammer; they even allow their infants to steal nuts from the mother's own collection, so that at times the great majority of nuts go to the infant. These actions probably serve to "scaffold" the infants' developing skill, and they clearly impose costs on the mothers. However, the infants gain immediately from the mother's actions (getting access to nuts, and to nut-meat with less effort from the mother than if she herself cracked them), so we can't be sure that the function of the mother's behavior was teaching rather than provisioning. Similarly, when studying the acquisition of the difficult skill of dealing with *Saba florida* fruit, Nadia Corp found that mothers shared their fruits with infants, which might have been argued to scaffold their learning; but donation was found to correlate with the time of life at which infants could digest the fruit, rather than when they most needed to learn about the task (Corp and Byrne 2002a). More compelling evidence comes from the extended monogamous families of tamarin monkeys (*Saguinus* and *Leontopithecus* spp.), where adults regularly donate food to infants and even have a specific call that signals their willingness to donate. Tamarin adults bias their giving toward foods that are novel for the infants, suggesting that promoting learning is the aim (Rapaport and Brown 2008). In an experimental study, the adults refused to share food as soon as the infants managed to solve the task independently; apparently, they really did track the progress of juvenile learning (Humle and Snowdon 2008).

Boesch also described two cases of what he called "active teaching" among the nut-cracking chimpanzees of Taï, giving convincing descriptions of mothers adjusting their behavior specifically to deal with inadequacies of their infant's technique. In one, the mother intervened just as her son had placed a nut wrongly on the anvil; she cleaned the anvil and repositioned the nut before allowing her son to proceed and crack it successfully. Another mother only intervened after several minutes of her daughter unsuccessfully struggling to crack nuts with an irregularly shaped hammer stone. When the mother approached, the daughter gave her the stone, which she then very slowly rotated into the correct position—taking a full minute to do so—and then proceeded to crack several nuts, most of which the daughter ate. When she left, the daughter adopted the same grip on the hammer as her mother, and managed to open the next few nuts. These two cases clearly suggest that the mother chimpanzees analyzed the nature of their infants' difficulties, and intervened specifically in a way adjusted to teach them the right way. The mystery, however, is why this sort of behavior, if it is indeed within the capacity of the chimpanzee, is not far more common. Nothing so convincing among wild chimpanzees has been described, even at the same site, in the many years since this work was published in 1992.

Some evidence for insightful teaching has also been found in elephants and orcas. Young female African elephants need to learn that, when they come into estrus, they can avoid the boisterous and often stressful attentions of adolescent males. The trick is to stay near an adult bull in musth; but these bulls are huge and must seem very frightening to young females, which often avoid them. Most work out how to behave on their own, but sometimes older female relatives "simulate" the characteristic behavioral signs of estrus while approaching the big bull, as if to teach the nulliparous female how to behave. The estrous state has a characteristic olfactory signature, so the big bulls won't be fooled for long into thinking the older female is in estrus herself (although researchers sometimes have been, only realizing their error when they checked long-term records of cycling and pregnancy!). By collating records of simulated estrus over 35 years, Lucy Bates and colleagues found that only 2 percent of estrous events were simulated (false), but these events were strongly associated with the first estrus of a younger female in the same family (Bates et al. 2010). Because most females are in states of pregnancy or lactation, finding two females in a group in estrus at the same time would normally be unusual. Naturally, the researchers checked to see if the "instructed" females had profited from the putative teaching: were they mated more by a musth bull, or did they become pregnant quicker than "uninstructed" females? The answer was no; but then, that is precisely what we should expect, if the teaching were targeted accurately at just those individuals who needed extra help and the teaching worked (Byrne and Rapaport 2011). A teacher who picks out those few children who are struggling with a concept the others find easy, and gives them extra help, would be delighted to find that they had reached the same level of competence as the rest. It remains puzzling that simulated estrus is not *always* associated with the estrous state of a relative in the same group: what other function does it have? Until that is clarified, these data are only suggestive of the possibility that elephants teach on the basis of insight into the understanding of others.

In several different places the local population of orcas (*Orcinus orca*) has developed a spectacular feeding technique of deliberate stranding in order to capture seals and sea lions from the edge of the beach. Orca calves in groups that rely on deliberate stranding need to learn how to escape from the beach, and presumably they have to learn that intentional stranding is not as dangerous as it looks. In the Crozet Islands, a mother orca has been observed to push her calf onto a beach, allowing it to grab onto an elephant seal, then pushing both her calf and the seal into the water (Guinet and Bouvier 1995). Even though the calf gained food in the process, it is hard to imagine that this could be sufficient motivation for carrying out such an apparently risky act with one's own offspring, whereas it is critical for Crozet orcas to learn the skill. It is also intriguing that, just as with elephants, the skill that appears to be taught is apparently a counterintuitive one for novices. However, as yet no systematic study has been published on these maternal interventions during the acquisition of intentional stranding by young orcas, so again we cannot be quite sure.

Summary

This chapter has reviewed evidence for a miscellany of rather disparate capacities, which nevertheless share the common feature that some sort of insight into individuals as sentient agents (including the self) seems to be necessary to explain them. Whereas in the last chapter we concluded that sophisticated perception, efficient memory, and rapid learning could explain all the phenomena discussed, here those things alone will not do. Empathy, in the sense of showing sympathetic concern and being able to take the other's perspective with regard to their feelings and needs, has been shown in chimpanzees, bonobos, gorillas, and elephants—and also, interestingly, in a corvid bird, the rook. Chimpanzees have shown that they understand the different role of the other in a two-role cooperative task, in contrast to monkeys who are able to learn either role but show no comprehension when the roles are suddenly reversed. Extensive experimentation into the ability to attribute knowledge vs. ignorance has given convincing evidence of this ability in chimpanzees and monkeys; and also in two corvid species, the scrub jay and the raven. Although naturalistic observations suggest that great apes can go further, and compute with false beliefs of other individuals, this has yet to be shown experimentally. Consistent with their ability to understand ignorance, chimpanzees also show some evidence of teaching based on understanding the specific learning needs of the other, as do elephants and orcas; however, in none of these cases is the evidence watertight. More difficult to interpret, though based on many years of careful experimentation, is the fact that all species of great apes, along with elephants, probably dolphins, and—again—a corvid bird, the magpie, are able to understand that a mirror reflection is of themselves. And increasing evidence is beginning to suggest that apes (or at least chimpanzees) and elephants have some understanding of the nature of death.

The coincidence of finding (most of) the same group of species in all these traits—great apes, toothed whales, corvids, and elephants—begs for a unitary explanation, despite the apparent variety of manifestations. Most of these abilities are treated in the literature in the context of theory of mind, but for some of them it is not difficult to construct explanations that require only a mechanical understanding of individuals as objects, not subjects. For instance, the role of the other in a cooperative task may be described in terms of what changes the partner can effect in the environment, not necessarily what they think or know; and to work out that the body reflected in a mirror is oneself is something that should be computable merely from cross-modal matching. Yet to do so misses the point, that the vast mass of animals apparently cannot do these things at all.

Suppose all these tasks require the ability *to represent individuals as agents with feelings and knowledge* (other individuals, and the self): in other words to mentalize about the situation. Then, a common strand may perhaps be perceived. With such ability, specific tests of others' knowledge or ignorance should naturally be passed. Recognizing ignorance in close kin or allies may inspire teaching. Recognizing the difficulties others are facing from their perspective may inspire sympathetic concern or empathic direction of help. Understanding the complementary roles of a cooperative task, in the sense of appreciating both roles when trained from the perspective of only one, would be helped by realizing how privileged knowledge may enable success. With the ability to mentally represent oneself as a living agent just like others of one's species, it should be possible to learn who "that puzzling image in the mirror" is, which moves in time with oneself. Finally, while being able to represent oneself as a living agent gives no guarantee of understanding death, it is certainly true that until one can understand life there is no possibility of representing an end of the state of living.

On this interpretation, the ability to mentalize—to represent oneself as a living agent, and to predict others' behavior by computing on the basis of some of their mental states—has evolved more than once. Four times in fact: once in the ancestor of great apes, since all living apes show (several of) these abilities, whereas the closest out-group, the Old World monkeys, generally do not; once in cetaceans, probably in the ancestor of the toothed whales, but possibly earlier—it is hard to be precise when we have so little evidence one way or the other about baleen whales or hippopotami; once in a proboscidean, since the living Asian and African elephants both show several of these abilities; and once in an early corvid ancestor, to judge from the scatter of special skills in ravens, rooks, jays, and magpies. What might have led to the repeated evolution of mentalizing ability? This is a question to which we will have to return in later chapters, but for the moment note only that all these groups are considered to have "unexpectedly large" brains.

One possibility is that mentalizing ability is an automatic spin-off from developing an (absolutely) large brain: once the brain exceeds some criterion, mentalizing becomes possible (Dunbar 2003). On this scenario, the evolutionary pressure for large (absolute) brain size may have been any number of things: social complexity, technical complexity, environmental complexity, or something yet undreamed of—or even none of these direct effects, if the large brain is just an allometric consequence of selection for very large body size, for example because of advantages in reducing predation.

An alternative possibility is that cognitive need drove brain change, either of quantitative or qualitative nature. The compelling advantages of being able to compute with mental states may have led to enlargement of certain brain areas, with consequent enlargement of the whole brain as an indirect effect. (We might also expect compensatory brain losses of other parts to have occurred.) And if selection acts to increase brain size, then we should expect subsequent body size enlargement to result. Fossil evidence may sometimes allow this sequence to be detected: did species evolve larger bodies before, at the same time as, or after their large brains?

Finally, we might consider a scenario of qualitative change: that mentalizing is a function of specific circuitry. In that case, the same or equivalent computing algorithms must have developed independently in four unrelated taxa. In humans, fMRI work has identified brain areas—the medial pre-frontal cortex, including the anterior cingulate cortex, and the inferior frontal gyrus—that are reliably active when we are involved in mentalizing, that is, processing of states of the self and of others in conjunction with emotional information (Keysers 2011). These same brain regions, along with the insula located just below them, an area known to be involved in emotions, have proved to contain spindle neurons (von Economo neurons), large cells that apparently enable communication across the brain and have been described as the "air traffic controllers of emotion" (Allman et al. 2001). Intriguingly, these neurons have also been found in the brains of great apes, elephants, and cetaceans (Allman et al. 2010; Hakeem et al. 2009). The coincidence suggests convergent evolution of neural machinery that may be basic to empathic and perhaps mentalizing abilities in these species—and specifically in large brains, explaining the association of large brain size and greater signs of empathy and mentalizing. It must be remembered, however, that at present we have little idea of the computational operations allowed by von Economo neurons, or even by particular cortical areas, so these associations allow no more than speculation.

Chapter 9

Pivot point

From social to technical abilities

Up to here, by taking the approach of seeking evidence of cognitive sophistication within the social and cultural realms, we have seen a number of indications of insightful understanding. Recall that what I'm meaning by "insight" is the ability to understand how things or people work, by means of computing with mental representations of them. Signs of insight have included an animal assessing how much another individual has understood its intentional signal; reacting appropriately to what others are currently able to see, or have been able to see recently and therefore know about; being able to take either role in a two-person cooperative task after learning only one of them by experience; deliberately planning to deceive others, or to help them learn by targeted teaching; understanding the image in a mirror as being of the self; and even some sort of understanding of death. These are mainly social insights: insight into technical intricacies has barely been touched upon. Only when it came to the ability to learn socially a complex and novel task did I raise the possibility that the underlying capacity for insight might depend on physical as well as social representations. The various social indications of insight were discovered in an eclectic scatter of species—crows, dolphins, elephants, and parrots—but among our closest relatives, the primates, the finger pointed time and time again at the great apes. It seems, therefore, that within our own lineage the capacity to show insight made a "quantum leap" when the common ancestor of the living great apes—orangutans, gorillas, chimpanzees, and humans—separated from the ancestor of today's monkeys, and this jump was toward social insight. What might have led to this change?

Ever-increasing social challenges?

A possible clue comes from the finding, noted already, that a species' brain size correlates with its social complexity—in the sense of the number of individually known companions that typically live in semi-permanent association and have to deal with each other in both competition and cooperation. The statistical association, between the average size of groups in which animals know each other as individuals, and the size of the whole brain or specifically the neocortex, has strongly supported the "social intelligence hypothesis"—that a powerful selection pressure for specialization in cognitive skills comes from the need to solve problems in the social realm.

In Nicholas Humphrey's seminal chapter (1976), which introduced the idea of social intelligence to a wide audience, he explicitly included understanding of others' social intentions within the broad category of intelligence. As we saw in Chapter 6, social intelligence theory succeeds in accounting for the large brains and impressive social abilities of simian primate species, as compared to (say) rodents. Monkeys and apes have been shown to recognize dozens of others as individuals, to react appropriately to their many different moods and intentions, to remember their past histories of interaction with the self, to work out their kinship and relative ranks, and so forth. For us, as humans, it is natural to mentally represent the array of social information involved in these abilities and compute with that array: our social sophistication is based on showing insight. However, that is not actually necessary for any of the social skills in that list: in all those cases, simpler evolved "rules" will do the trick, and there is no evidence—for monkeys—that any more is involved. But perhaps social complexity can nevertheless be extended to account for the evolution of more insightful understanding in great apes?

Unfortunately for this idea, great apes simply don't live in particularly large semi-permanent groups. Orangutans are essentially solitary, with the only long-lasting association that between mother and dependent children; the Sumatran species is more often seen in small groups than the Bornean one, but even these are mere temporary gatherings found when food is locally abundant. Their interactions are perhaps best likened to how we are with neighbors in a city: we may know who they are and nod when we pass, but we don't know much about them or go out of our way to find out. Gorillas travel in social groups, averaging around 10 individuals in the western gorilla, sometimes up to 50 in the eastern, but usually less. Bonobos have been less widely studied, but also typically live in groups between 10 and 50 in size. Only the chimpanzee is regularly found in communities much larger than this, with group sizes of 80+ known from several areas of Africa. None of these figures would be at all exceptional among monkeys: macaques, baboons, squirrel monkeys, and talapoin monkeys have all frequently been recorded living in larger groups than any of them.

In an effort to stretch the social intelligence theory to explain the qualitatively greater abilities of great apes, Robin Dunbar has suggested that the critical extra challenge for great apes is that they form "fission–fusion" societies, in which parties are continually splitting and joining, rather than traveling always as a cohesive group. Chimpanzees and bonobos certainly do so; gorillas usually don't, yet evidence of insightful understanding occurs in them, too. Orangutans have long intrigued scientists, since they behave as if they were social species when they meet: when plentiful food brings them together they show behavior that is quite unlike the guarded tolerance of others shown by a genuinely solitary species like the brown bear. Indeed, in zoos, orangutans are successfully kept in social groups, and seem at least as content as similarly housed chimpanzees. Yet in the wild, orangutans seldom meet, and it has not been possible to make objective estimates of their "real" group size—the set of others an orangutan distinguishes as individuals with distinct characters and personal histories. Rather, a notional group size has been computed from their brain size, which becomes rather circular. Even worse for any attempt to use fission–fusion ranging to explain why apes are different to monkeys in how social intelligence theory applies to them is the fact that some monkeys are also organized in fission–fusion societies. Most spider monkey species show fission–fusion ranging, for instance, yet they have not been noticed to show special signs of insight. Only the belief that great apes are "cleverer" leads researchers sometimes to suggest that apes' social lives are intrinsically more complex than those of monkeys; to go on and use this difference in social complexity to explain apes' greater intelligence would again be circular logic.

There seems no principled way of getting around the fact that, on social intelligence theory, monkeys and great apes should overlap each other in signs of insightful intelligence—yet they do not. Great apes do, however, differ systematically from monkeys in one aspect that may be the key to their different mentality. In a variety of ways great apes show much greater technical skills than monkeys, in gathering food from sources that are nutritious but present difficulties.

Clever ways of feeding?

We have already met some ways in which great apes show technical ability in their foraging, in Chapter 7, when we looked at evidence for cultural transmission of skill. As we saw, chimpanzees use tools to feed on insects that are hidden and protected by nests, by "fishing" into narrow holes with flexible probes, or use longer, rigid wands to "dip" into bivouacs of the aggressive African safari ant. An understanding of the relation between tool form and function is particularly clearly shown by cases where chimpanzees use more than one type of tool successively, in the pursuit of a single goal: a suite of tools. Chimpanzees at Ndoki (Congo) regularly do this, employing strong, pointed twigs to force entry to termite nests, then using thin, flexible herb stems to extract the termites

(Suzuki et al. 1995); and Brewer and McGrew (1990) described a chimpanzee successively employing four sticks of differing thickness and length to get the honey in a nest of wild bees: a strong tool to break a hole in the casing, a slender probe to investigate inside, and so on. Most remarkable of all, at Goualougo (Congo), the chimpanzees have been described to use sets of two different tools for no less than 11 different feeding tasks (Sanz and Morgan 2009; Sanz et al. 2009). To eat terrestrial termites, for instance, a Goualougo chimpanzee first uses a strong, smooth rod to probe deeply into the forest floor, forcing it down with considerable power, each time retracting the tool and sniffing the end until it can scent a pierced termite nest. Then, the rod is discarded, and a flexible, slender stem, whose end has first been frayed into a "brush" by pulling over the teeth, is threaded into the hole created by the rod, then carefully retrieved. As with the more familiar mound-termite species, aggressive soldier insects attack the chimpanzee's tool and can be pulled out to eat. Intriguingly, chimpanzees arriving at suitable sites for subterranean termite fishing usually come with tools ready prepared—but only of the second type, the brush-tipped flexible stem. The explanation is that the first sort, the strong wands, can be re-used many times and there are usually a few lying around, whereas the delicate brush-tipped probes need to be made anew each time. Evidently, the chimpanzees are able to anticipate their technical needs in advance (Byrne et al. 2013). We do the same when we mentally review planned future operations to work out what we'll need: think of deciding the shopping list to buy in advance of cooking a new dish in the evening. The tool-using skills of chimpanzees not only show an understanding of objects as tools, but they also require delicate coordination of the two hands, and in some cases more than one stage of processing must be sequenced correctly, a sequence that can be anticipated in advance for planning purposes.

One possibility, then, is that an increasing need for sophisticated tool use selected directly for great ape intelligence. This theory must confront the puzzle that all great apes in captivity show tool-making skill, yet only the common chimpanzee and one population of orangutans do so with regularity under natural conditions (McGrew 1989). Several researchers have consequently speculated that tool making may have been found in the common ancestor of living great apes, but has since been "given up" by modern gorillas, bonobos, and most populations of orangutans. If so, then whatever environmental challenges selected for tool making in that ancestral species might have led to the cognitive advance shown in all its descendants. It is rather unsatisfying, however, to rely upon an explanation which proposes that a trait was derived in the common ancestor of a clade, yet is now shown by only two species out of six in it.

A version of this hypothesis highlights the concealed and hard-to-extract nature of the food itself. Susan Parker and Kathleen Gibson (1977, 1979) have pointed out that species that obtain hidden food resources, yet lack an appropriately specialized anatomy, have a problem. Where the need to forage by extraction is year-round, specialized foraging mechanisms might be expected: such as an aye-aye's middle finger, a woodpecker's long tongue, and also in the case of the gorilla, which they suggest "pulls tubers from the ground with his massive strength" (Gibson 1986). However, where the need arises seasonally and over a wide range of foods, "sensorimotor intelligence" would be favored and intelligent tool use will result (Gibson 1986, p. 99; Parker 2015; Parker and Gibson 1977, p. 634). They proposed that this was the case for the ancestral great ape (which they suggested to resemble the modern chimpanzee in its ecology), and independently in the capuchin monkey. The recent discovery that some populations of the bearded capuchin (*Cebus libidinosus*) regularly crack nuts by selecting and carrying suitable stones to use as hammers on stone anvils, very much in the manner of West African chimpanzees (Fragaszy et al. 2004; Moura and Lee 2004; Visalberghi et al. 2009), has lent support to this *extractive foraging hypothesis*. But there are difficulties. First, as noted, there is the unsatisfying nature of proposing that the common ancestor of all great apes was a tool user. A second problem is that the tool use of capuchins appears to be of a non-representational, non-insightful sort (Sabbatini et al. 2012; Visalberghi and Limongelli 1994). It could be argued that capuchins just happened not to possess the genetic variation necessary to develop in the way that ancestral great apes did. Evolution is always limited by the genetic variation available, but it's a bit of a stretch to propose this as an explanation with species as closely related as apes and monkeys. Indeed, the generality of extractive foraging among other mammals and birds has been used to question the whole idea that it was a significant influence in the development of primate intelligence (King 1986).

Understanding the significance of great ape tool use has been impeded, I believe, because the tool as object has dominated researchers' attention, rather than the manner in which it is employed. Consider the way chimpanzees sometimes employ tools: it suggests considerable ability at planning organized motor movements of the hands, and just that sort of ability may in principle be shown in other ways of foraging as well. With this perspective, let's re-examine the feeding behavior of apes other than the chimpanzee, to see if they too show organized manual skills—skills that perhaps happen not to benefit from incorporating tool use.

On current evidence, the laurels for complex organization of feeding sequences go to the mountain gorillas of the Virunga Volcanoes in Rwanda (Byrne 1996;

Byrne and Byrne 1991, 1993). The major diet items available to this population are nutritious and digestible, but all are physically defended or enclosed in some way—by spines, stings, hard casing, or tiny clinging hooks. Each of these challenges requires a very different technique to deal with it, and each technique involves several sequential stages. Throughout the several stages the two hands are used together, in different but synchronized roles. Sub-sequences (of various numbers of steps) can be repeated iteratively until some criterion (a sub-goal) is reached, whereupon the main sequence continues. Thus, the sub-sequences are treated as sub-routines. These characteristics show that the gorillas' mental programs are organized hierarchically, not in the linear chains of actions that would be expected from associative learning. Nevertheless, juveniles attain adult levels of efficiency by the time they are weaned at 3–4 years old. Unlike the skilled tool use of chimpanzees and orangutans, which is employed only at certain times and much more by some individuals than others, the multi-stage hierarchical programs of mountain gorilla food preparation are used every day by every individual, because they are essential for consuming the major foods. For such a large mammal, with a simple gut living in a cold environment, speed and efficiency in processing must be at a premium, and the techniques used by the gorilla appear optimal for the job.

Orangutans, particularly of the Bornean species which lives in dipterocarp forests on impoverished soils, forests known for erratic mast-fruiting, may face similarly severe tests of skill in dealing with hard, physically defended fruit (Galdikas and Vasey 1992). Anne Russon has analyzed orangutan feeding techniques, and shown that they are strongly reminiscent of those of gorillas in their complexity, but here involve multi-stage programs of action to reach a safe and convenient platform, in addition to the manipulations of the foodstuffs themselves (Russon 1998).

Thus all three genera of non-human great ape—*Pan*, *Pongo*, and *Gorilla*—show an ability to construct complex, highly organized programs of action that are tailored to their own specific foraging problems. To build elaborate but novel programs of manual actions, whether for plant or insect gathering and whether involving tools or not, great apes rely on skills that differ qualitatively from anything described in monkeys. Why should the great apes, alone among the primates, have evolved these qualitatively special abilities? There are several possible reasons that together point to what is likely to have happened in great ape evolution.

An evolutionary hypothesis: Smarter food acquisition

In almost all the forests they inhabit in the Old World, African or Asian, apes today must compete with monkeys, often of several different species. Those monkeys have two big advantages; one is simply that they are quadrupeds, like most mammals. Apes are brachiators. Brachiation is the kind of locomotion that allows gibbons to swing dramatically through the treetops, but its primary evolutionary advantage is in foraging: apes can hang below branches, thereby gaining convenient access to fruits that are hard to reach if you are restricted to walking on top of branches, as a quadruped is. The suite of skeletal and muscular adaptations that allows brachiation includes arms much longer than legs, and a shoulder blade that moves freely over the rib cage, thus allowing the arm to pivot above the head. Brachiation comes with costs, however: it makes quadrupedal locomotion on a flat surface difficult. Try it: as humans, we share the adaptations for brachiation that date from shared ancestry with other ape species, although in our case the more recent adaptations for bipedal walking (long, powerful legs; narrow hips) make it even harder for us to walk quadrupedally. The apes tackle this problem in various ways, from waddling bipedalism (gibbons) to specialized knuckle-walking (chimpanzees), but none is as efficient as quadrupedal walking. For gibbons, with specialized rapid brachiation and a small body size that enables them to live year-round in small territories, there's no great problem. The great apes, on the other hand, are just that: great. They are big and heavy, and so they need to travel over a wide range to feed; the various ways apes can travel are inefficient when moving over long distances, compared to quadrupedal walking which apes can't do. The monkeys' second advantage is dietary. Cercopithecine monkeys—the major food-competitor species for apes in the Old World forests where they live—benefit from digestive systems able to deal with coarser leaves and less ripe fruit than can those of apes. Other things being equal, larger animals can digest coarser foods: large deer can eat tougher plants than small deer, and so on. But among the primates, unusually, Cercopithecine monkeys happen to have guts specialized for digestion of rather coarse plant material and rather unripe fruit, whereas the guts of the larger apes are less specialized. Cercopithecine monkeys can therefore exploit foods before they become of value to apes; they even have cheek pouches that enable them to ingest extra food rapidly for chewing later, away from any competition. And representatives of those monkeys are found in virtually all habitats where apes live. That means a great ape is up against the worst kind of food competition: from smaller, faster, and energetically more efficient animals that can eat its preferred fruit before the ape can, and when times are hard can fall back on tough plant foods that the ape can't digest. The puzzle is really why the great apes are not extinct. How do they survive in the face of monkey competition?

When I first discussed this puzzle with paleontologist friends, their response surprised me: "Well, actually most apes are extinct!" Today, even when *Pongo,*

Gorilla, and *Pan* are recognized each to include two species, something that could not be known from fossil evidence, there are a paltry six great ape species, plus ourselves; but during the Miocene, from 23 Mya to 5 Mya, there were over a hundred species. Almost all their lineages are now extinct. When great apes flourished in the Miocene it was much warmer than the present day; but gradually the climate became cooler and dryer as a result of tectonic changes, including the uplift of East Africa and the increasing height of the Himalaya in Asia. As seasonality increased, the great rainforests that then extended over much of Europe, North Africa, and China receded toward the much more restricted forests of today. More open, arid habitats prevailed in many areas. Apparently as a direct result of these ecological changes, the apes began to die out. Clearly, however, not all apes became extinct at that time: what did the surviving species have that enabled them to survive in the face of devastating climate change? I suggest that the answer is that they developed a new sort of cognition, enabling them to get to a wider range of foods—foods that did not seasonally vanish as ripe fruits do in a monsoon or temperate climate.

The hypothesis, then, is that the cooler, dryer climate set up a strong selection pressure for the ability to feed in more seasonal environments. That resulted—in those few ape species which survived, the ancestors of living great apes—in the development of skilled feeding techniques, some incorporating tool use, some not, but all based on an ability to build organized programs of coordinated actions. In today's forests, this also enables the surviving apes to deal with competition from monkeys that superficially appears devastating, by allowing them to get to foods which monkeys cannot reach, using skills that are technically demanding compared to those of monkeys. We can see this happening. Monkeys love eating termites when they emerge from the mound in mating flights, but that happens on only one day of the year for each termite species. Chimpanzees make and prepare fishing tools from natural materials, and operate them skillfully to get access to termites inside rock-hard mounds or deep under the forest floor, so they can eat termites over a much longer period. Using similar techniques, they have access to arboreal *Campanotus* ants year-round. *Dorylus* ants are present on the surface year-round in Africa, because their powerfully biting soldiers are so formidable that they have nothing to fear—from monkeys. But chimpanzees use a different technique of skillful foraging—ant fishing—to eat them with some safety. *Coula* and especially *Panda* nuts have an exceptionally hard outer casing: most animals, including monkeys, cannot bite into their nutritious centers, so the nuts lie around on the surface for long periods. Chimpanzees, however, have developed the skill of hammer-and-anvil use, using stones or hard wooden clubs to break the nuts open. Shown most clearly in gorillas, skilled feeding gives access to the soft pith inside a woody casing, and to nutritious leaves, available year-round but unpleasant to eat because of spines or stings. For orangutans, living in forests with dramatic periods of ripe-fruit shortage, fruits growing on rattan and palm trees are a potential fallback: safe from eating by monkeys because the plants are defended by vicious spines that prevent direct climbing. Orangutans develop skillful indirect routes by which they can clamber delicately to positions where the fruits can be eaten safely: often a small tree needs to be set in motion by body swinging in order to cross particularly large gaps. Honey is present year-round, but the bees nest inside tree trunks: orangutans use probe tools to get it out.

The talent that underlies all these "clever" feeding abilities is the ability to develop programs of behavior, in which many simple actions of each hand (and feet, in the case of orangutans!) are coordinated into an organization tailored to achieve a particular goal. The remaining chapters will develop the hypothesis that this cognitive advance gave special abilities to the apes when it comes to interpreting and constructing plans of action, involving representation of actions of both the self and others. This representational ability included physical objects, allowing them to be specified in an abstract way as potential tools. From mental representation of objects and actions, I will argue, planning evolved to include other individuals as active agents with intentions. If this hypothesis is correct, the old phrase "food for thought" has perhaps a literal meaning: the ape's need to feed more skillfully may have led directly to the thinking skills in which we as humans exult. Since developing this theory requires a switch away from social cognition to the world of objects, we will begin by examining whether, in primates or any other species of animal, there are signs of insight in dealing with object-knowledge.

Chapter 10
Knowledge about the physical world

What animals know about the physical world has been reviewed in depth many times before (e.g., see the excellent treatment in chapters 2–5 of Tomasello and Call 1997), and it is not the aim of this chapter to offer another summary. Instead, I will focus selectively on behavior that might indicate that individuals have insight into how things in the world work; I'll therefore skim over most of the immense amount of work on the physical cognition of animals. Although many clever puzzles have been posed to various species of animal in order to test their abilities in laboratories, my emphasis here will be on real-life situations: finding remembered items when they are needed, minimizing effort when exploiting a home range, processing difficult foods, or reducing the risks of predation.

Memory of places: Where, what, and when

Foraging animals of many species show every sign of using their prior experience to search for resources in a well-organized way: it therefore often appears to observers that they must have remembered where things are located, rather than just searching "cold." Indeed, a prominent approach to understanding foraging in modern biology, optimal foraging theory, takes it as a fundamental premise that all animals have evolved to use their environment in the best possible way for their particular needs, and uses this assumption to evaluate whether researchers have properly understood those needs. Do animals, then, have long-term memory of what is where?

Some of the most impressive tests of that possibility have been carried out with food-storing birds which, like squirrels, cache food when it is abundant in order to retrieve it later: jays, nutcrackers, and various species of tit or chickadee. For example, Clark's nutcrackers cache up to 30,000 nuts each autumn which they retrieve over the next six months (Balda and Kamil 1992). All the evidence so far suggests food-storing birds do remember the specific places in which they stored some and perhaps all of the items, rather than just searching every location of the general type in which they always tend to store foods. In captivity, food-storing birds have been found accurate at retrieving large numbers of items they stored (for review, see Shettleworth 1998). Two such species, the western scrub jay and the black-capped chickadee, have also shown some understanding of how different foods change over time. Birds were allowed to cache some items which perish rapidly over time, like insect larvae, and others that do not, like peanuts (Clayton and Dickinson 1998; Feeney et al. 2009). After a short delay, the birds retrieved the more desirable larvae, but when there was a longer delay before testing they abandoned the larvae and retrieved the long-lasting nuts instead. Thus, both these bird species can remember what they cached, where, and approximately when: a *what–where–when* or *episodic-like* memory. (These terms are used to circumvent unproductive argument about whether birds have "episodic memory," which would also include having the mental experience of being at locations now remote in time and space: Tulving 1972, 2002—something clearly untestable without a conversation.)

Rather similar memory abilities have been shown in monkeys, by means of detailed field study of their foraging behavior, so there is no reason to think these skills are unique to food-storing birds. In the dry season, baboons of the high veldt of South Africa rely on fruiting fig trees for their prime diet items; however, competition for these resources is fierce, so it makes sense to arrive early to exploit those fruits newly ripened overnight. Small troops are unable to defend prized resources from larger ones, so they are particularly vulnerable to competition: the movements of one such troop were studied in detail by Rahel Noser (Noser and Byrne 2007b). In the dry season, the baboons left their safe sleeping site early and traveled in a straight line to fig trees, over distances much greater than the trees themselves or any cue to their presence could be seen. In doing so, they bypassed many alternative sources of food, but returned to eat those later the same day. Why pass by good food? Because those foods, unlike the figs, were locally abundant, or scattered and slow to harvest, so were not vulnerable to competition. Just like an efficient shopper in the New Year sales, who bypasses the department store's food section and popular café to head straight for the bargains, coming back later to do their essential weekly shopping and have a coffee, the baboons were able to use knowledge of what is where, and how long they had to exploit it, to plan their days. It seems that, like scrub jays and black-capped chickadees, baboons have excellent time-dated memories. Noser tested the flexibility of baboon spatial memory by setting up five artificial provisioning sites, out of sight of each other, containing high-priority food: and she set them out at a time when the single male of the group was also concerned to protect his group from extra-troop incursions, giving him a dilemma (Noser and Byrne 2015). The male repeatedly broke off travel with the group to exploit the feeding sites, visiting them in several different sequences, interrupting his feeding at virtually any point to return to the group. When he returned, he was able to avoid revisiting depleted sites, apparently based on his what–where–when memory of the previous foraging episode.

A study of mangabeys, forest relatives of baboons, shows that monkeys not only remember what is where, but can also take the time-course of natural processes into account. Mangabeys feed on both ripe fruit and the insect larvae that sometimes infest the fruit. They often visit trees of the right type but which happen to yield neither resource that day: when this occurs, do they take note for future reference of any nearly ripe fruit or increasing insect infestations? Karline Janmaat followed mangabey troops for many days in a row to find out. When the monkeys next came close to a tree they'd visited recently, she recorded whether they went out of their way to check it again or just passed it by (Janmaat et al. 2006). By correlating their behavior with weather observations, she showed that the mangabeys were more likely to revisit if the weather had been warm and sunny on the intervening days, provided the tree had previously shown food that was nearly ready to exploit. Like a fruit farmer, the monkeys took account of the weather's effect on ripening. Crucially, the effect of intervening weather was just as clear if the resource was insect larvae as it was with ripe fruit. While ripeness can be detected by smell, and so a huge mass of ripe fruit might in principle be detected at long range, insect larvae have virtually no scent. Thus, the mangabeys' behavior shows that they can remember the state as well as the location of a tree they last visited several days ago, and they also notice and recall what the weather has been like since then. Impressively, they can put these memories together to simulate what has likely been happening to the fruit in their absence. Evidently their mental representations include some understanding of the ripening of fruit and maturing of insect larvae, enabling them to compute new information about the out-of-sight trees from their what–where–when memories of their own past explorations.

Judging by the diversity of species that have shown the use of what–where–when memories, the ability may be very general in animals—or at least in species of animal likely to benefit from them. Scrub jays are corvids, a large-brained group notable for impressive performance in other domains; chickadees, however, are not. In many spheres, as we have seen, it is the large-brained apes which have shown signs of insight, not monkeys: but baboons and mangabeys are monkeys. The best bet (put in more scientific terms, the appropriate working hypothesis) is that a wide range of animal species remember where useful things are located and decide when to go back to them on the basis of what may have happened to them since they were last there. The next category of behavior examined is similarly likely to be based on rather general abilities.

Planning efficient navigation in large-scale space

The phrase "large-scale space" is important: if an area can be viewed from a single location, then picking a route through it may be done without use of memory and it would be inappropriate to use the term planning for that exercise. The tiny predatory jumping spiders, of genus *Portia*, may be an interesting exception to that generalization (Wilcox and Jackson 2002). *Portia* spiders are able to decide and execute a route to a point directly above their prey, involving travel in a detour over a distance of many times their body length, in order to drop down on the prey undetected. The route chosen by the jumping spider is efficient and avoids dead ends. Jumping spiders have exceptional eyesight, for spiders, but whether *Portia* spiders are able to see in advance the whole layout over which they travel is not known.

If an animal decides in advance to take a certain route through the environment, going to places that are out of view at the start and will take significant time to reach, then it must be using some sort of mental representation: often called a *cognitive map*. However, travel need not be planned in advance to be efficient. Consider a fruit-loving primate, who happens to have no mental map. Using the simple strategy of generally going in a straight line, which avoids searching areas already exploited, picking up information from indirect cues, such as the sound of dropping fruit or the calls of fruit-eating birds, as well as direct cues like the scent plume of a big fruiting tree, and stopping only when it gets to a fine display of fruit, produces a result that—when the route is plotted—will look like bee-line travel to an obviously valuable resource, just as if the route had been planned in advance (Byrne 2000a). Researchers have to be careful!

"Cognitive map" is also an ambiguous term, having been used in two distinct senses. Cognitive psychologists use it to imply any kind of stored information that allows effective planning of travel; researchers on bees and birds, however, have used cognitive map to imply stored information that corresponds in Euclidian terms to the spatial structure of the environment, as does a printed map bought from a map shop. Here, I will use the broader sense of cognitive map, any mental representation that allows travel planning; but where possible I'll also examine the evidence for what kind of map, Euclidian or otherwise.

The behavior described in the last section shows that baboons and mangabey monkeys use cognitive maps to make their travel more efficient, and this has been confirmed for other primate species, such as orangutans (Mackinnon 1978), chimpanzees (Ban et al. 2014; Boesch and Boesch 1984; Janmaat et al. 2013), spider and howler monkeys (Di Fiore and Suarez 2007), saki monkeys (Cunningham and Janson 2007), tamarin monkeys (Garber 1988), and mouse lemurs (Joly and Zimmermann 2011): a list that covers all the major divisions of the primates. Leo Polanski and colleagues used the speeding-up of movement that initiated rapid travel direct to waterholes to discover how far African elephants anticipated ahead when they planned their route, and showed that these travel movements began typically 4.5 km from the water, sometimes up to 49 km, suggesting the use of detailed large-scale cognitive maps (Polansky et al. 2015). The clearest laboratory demonstration of cognitive maps, in the rat, relies on the use of the water maze (Morris 1981). In a water maze, cloudy water makes it impossible to see a platform just below the surface in a large circular pool; although rats are excellent swimmers, after a while they need to rest, so when they are dropped in the pool they swim about until they come upon it. After several experiences, the rat can swim directly to the platform, even though there are no local cues to its position. To show that flying animals use memory in travel is more challenging, but there is little doubt that—for instance—birds and bees can also decide on travel routes on the basis of past experience.

Animals may be quite limited in how far ahead they can plan with cognitive map knowledge. In general, experimental studies with monkeys have shown that they reliably choose the single best location to go to next, but they don't usually seem to plan more than one step ahead, let alone give any sign of computation with larger spatial units (Janson 2000). However, capuchin monkeys were able to compute an efficient two-location route: given the choice of two distant food sites, where the much better of the two was also a lot further away, they reliably visited the poorer resource on the way if the detour was not too great (Janson 2007). Pioneering studies with hamadryas baboons suggested that they "looked ahead" much further than that. When groups of hamadryas baboons leave the cliffs on which they have slept safely, they mill around in a loose mob before setting off to forage. Their movements have been likened to watching an amoeba under the microscope: "pseudopods" of baboons seem to flow out in one direction, only to falter and come back, repeatedly. Finally, one direction is adopted by the majority and the whole group pours out in that single direction, and then begins to fragment into the tiny units that forage alone. Researchers showed that the direction of the exodus, but not the direction of any subsequent travel, predicted where the baboons would later reunite into a group in the middle of the day, often a shady waterhole in the desert habitat (Sigg and Stolba 1981): as if the whole business of "pseudopods" was a series of proposals, a negotiation. However, this exciting possibility has never been confirmed, and the researchers did not consider that the baboons might be able to see the tall trees that often indicate the presence of water from the cliffs on which they slept, thus perhaps needing no cognitive map at all.

Whether animal travel is ever based on mental maps with Euclidian properties is another matter. One particularly telling clue would be if they traveled on a completely new route, just as we can when using a printed map. Experiments have been conducted with bees to find out if they can compute a novel shortcut, between two locations that have each been visited but never on the same route: the conclusion was a resounding "no" (Dyer 1991). Surprisingly, a similar lack of Euclidian knowledge has been found with rats in the water maze. When barriers were placed into the maze, so that not all of the distant landmarks around the experimental room were visible from every point, it was found that rats were only able to learn to swim to the platform if the panorama visible when dropped into the maze included part of the view visible when safely sitting on the platform (Benhamou 1996). Rather than developing a cognitive map with Euclidian spatial properties, it seems the rats learn what the view was like when they got to the right place: then, when placed in the pool on test, they swim to maximize the similarity between that and what they can currently see.

In the field, it is very difficult to know whether an animal is taking a novel shortcut since you don't normally know where it has previously traveled when you were not watching. The baboons studied by Rahel Noser, mentioned already, offered a novel way of tackling this problem. Because their travel in the dry season consisted mainly of long, straight-line travel segments that ended with obviously valuable goals, much of the time Noser could reliably predict where the group was heading. And because the group she studied was so small, when it met or even heard another baboon group its travel was usually interrupted. This presented a natural experiment (Noser and Byrne 2007a): when movement along their original travel route is disrupted, how will the baboons behave? If all their travel was on familiar routes, they might need to return to the previous route; but if their mental maps were Euclidian, they could simply take a bee-line route from the place they ended up after the encounter. Noser found no evidence for Euclidian map ability: every time, the new route was either along the same track as the old, or along another route they had traveled many times before and which happened to go the same way.

Currently, there is no evidence for animals using cognitive maps that have Euclidian properties, as do printed maps. On the other hand, it has to be mentioned that in familiar urban environments, there is little sign that we ourselves do any different. Many years ago, I studied these abilities in people. Residents highly familiar with their town consistently failed to demonstrate that they had any idea of the angles between major routes, even when they were very far from 90° (Byrne 1979). Consistent with these errors, when they tried to draw the familiar configuration of roads in the town center, they made each road meet the next at a 90° angle: after drawing a few roads, they soon had to admit, "Er, these two roads are supposed to be the same one, and it's really one straight road, but it's not straight on my map . . . !" Nor did they seem to know the distances between well-known places, tending instead to estimate distance based on what was memorable about the route (Byrne 1979). If a route had many notable locations—cafés, bars, shops—or many twists and turns, its length was greatly overestimated compared with a straight route through a monotonous housing area. Subjects reported that they were always surprised how long those straight, monotonous routes were when they had to walk them! It seems humans are typical primates, after all: our navigation in the everyday world is generally not based on Euclidian representations of space.

If a cognitive map does not have Euclidian properties, how can it be useful? Researchers often refer to "route-based" knowledge, and for non-human animals all knowledge of large-scale space must originally be derived from traveling along routes (as it also usually is when a child builds up knowledge of their local town or city). The simplest mental representation that could allow retracing of a route would be string-like, linking representations of distinctive places or landmarks in an order that was isomorphic to the real order of locations along the route. The Euclidian distance, even along the route, would not necessarily be preserved. This kind of representation is called a *route-map*; the "rutters" of medieval mariners were route-maps, listing a series of landmarks and which course to take to reach the next, painstakingly recorded by a pioneering navigator to give to others. We know, however, that animals are not constrained to travel only between pairs of locations they have visited on a single route before, so route-map representation is insufficient as a theory.

Imagine developing your experience of a city by exploring along routes from one place to another. Sooner or later, some of the routes will cross each other, and if you notice that you've reached the same spot from a different direction, immediately you start to have choices available. That point becomes a potential crossroads of routes you know can be traveled. In terms of mental representation, then, allow that the string-like representations of each route can become linked at "nodes" corresponding to the crossover points, forming a network instead of a string. If a *network-map* of this kind has to be forced onto a sheet of paper for display, the result would be like the famous London Underground map. (To fit the London Underground map on paper, and to look attractive, notional distances have been invented.) Think of the relationships between stations on the Underground map: they are correct, in terms of which comes next, but the distances between them are not preserved. From any station (node) you can trace a path to any other, hopping between Underground lines at interconnecting stations, but since distances and directions are not represented, the only way of picking an efficient route is to minimize the number of interconnecting stations. (It's easy to forget that the printed map of the London Underground system does not represent distance or directions, but if you try to use it as a map on the surface you soon hit evidence. I well remember my own dismay, after a long and roundabout Underground journey, noticing that the place I started from was actually in view when I surfaced at my destination. It was at the end of the street, a few hundred meters away!)

I think that on current evidence, a network-map is the best candidate for understanding the cognitive map abilities of rats, non-human primates, and indeed humans, when not actually clutching a paper map or GPS. When an individual of any of these species personally develops a rich and detailed set of travel knowledge about the large-scale space in which they travel, the knowledge is originally remembered in a route-map format, since that is how it is acquired; but with time, nodes develop between route-maps and eventually an interconnected network-map can be used to make travel decisions more generally. In the case of a primate like a baboon, living all its life in a few square kilometers of woodland, the number of interconnecting nodes and alternative pathways must be immense, dwarfing the complexity of the London Underground map. Travel is not restricted to routes experienced before in their entirety, since route hopping at nodes is possible, and the network-map can be searched for the shortest pathway between any pair of remembered locations. Given a richly interconnected network-map, distances will be minimized pretty efficiently, but the pathway will always be composed of segments traveled before—there will be no novel shortcuts. Thus, a researcher following the monkey, carrying a printed map or GPS, would sometimes be able to find a slightly more efficient route than the monkey takes. Of course, not generally knowing the monkey's goals, he won't often realize that. Only experiments like those I did with people, or the natural experiments Noser carried out with baboons, are able to show up the slight biases and minor failings of travel with a network-map representation. Moreover, the restriction that one must keep to familiar routes also has the virtue that nothing too unexpected is likely to intervene. We all remember cases where our travel planning in a new city with our nicely Euclidian paper maps came unstuck when two "intersecting" routes proved to be at different levels, the roads lacked pavements altogether, or were inexplicably blocked to pedestrians!

Tool use and understanding causes

Using tools is surprisingly widespread among animals (Shumaker et al. 2011). In most cases, the tool use is restricted to a single circumstance, in which a single type of tool is used for a single purpose. Often, indeed, all members of the species show the same kind of tool use under all appropriate conditions, suggesting that the development of the trait is strongly channeled by genes. For example, all hermit-crabs use a discarded mollusk shell as a refuge. Here, the "tool" is picked up rather than made, but some decision-making is required. Each time a crab grows too large for its current shell, it must select a shell that is larger—but not too large, or it will be vulnerable to predation. Only in a much smaller number of cases of animal tool-use is the tool made or prepared in some way: an example is the Galapagos woodpecker finch's use of cactus spines for probing. There are no woodpeckers on the Galapagos Islands, but a small finch has managed to exploit the niche usually filled by woodpeckers, using detached cactus spines as probes to get grubs out from underneath loose bark. The spines are usually pulled off cactus trees and carried to the site of use. Only lowland populations of the finch show the tool-using habit, probably because there is no advantage to using it in the wet forest high on Galapagos. Laboratory studies have shown that social learning is not involved: given the right circumstances, hand-reared finches learn to make and use probes perfectly well without ever seeing another finch doing so or finding any of their probes (Tebbich et al. 2001).

In other cases, social learning certainly is involved: an example is the use of anvils and hammers by the California sea otter. The sea otters—of California, but not those of the same species in Alaska—use stones as tools. An otter will swim down to the bottom, and use its prehensile front paws to pick up an abalone shell and a large stone. Returning to the surface, it swims on its back and puts the stone on its broad stomach, then smashes the shell down on the stone repeatedly until it breaks and it can access the mollusk inside. Another otter might pick up a hard-shelled crab when foraging in rock pools, keeping hold of the crab as it also picks up a sharp stone in the other front paw. Then, holding the crab in one paw and the sharp stone in the other, it will hit the crab shell until it breaks. Although hammering is common to both these skills, the actions are different and the appropriate tool is different: there are two tool-using abilities involved. However, no one otter ever shows both skills: daughters tend to match their mothers both in preference for foods (crab vs. abalone) and tools (small sharp stone vs. large flat stone). This shows the importance of social learning for acquiring the skills; but a single otter pup does not acquire a toolkit of two methods. It's either going to become a crab-cracker or an abalone-smasher (Riedman et al. 1989).

Against this general background, the behavior of wild chimpanzees is special. At most of the locations this species has been studied, right across Africa, several different kinds of tool are used by most or all members of a community, and the mix of methods varies from site to site. Each chimpanzee has a "toolkit," a repertoire of methods that depends on using objects as tools. As we saw in Chapter 7, the distribution of each distinctive kind of tool use between communities is better explained by ecological needs than by constraints on the spread of knowledge by social learning. Social learning may be important, however, for an individual chimpanzee's development of its tool repertoire. Indeed, some of the tool-using methods are so complicated and specific that it beggars belief that every individual could stumble upon just the same technique, or be shaped toward this norm by the constraints and affordances of the environment. Consider, for instance, the method by which Goualougo chimpanzees hunt terrestrial termites, described earlier. How could a chimpanzee even know that there was a nest of termites 0.5–1 m below the forest floor in a particular place, if it had not seen other chimpanzees extracting them? To succeed in getting a decent termite meal, the chimpanzee must select or make two different tools with different properties, and use them in sequence: a long, smooth, robust rod and a slender, flexible probe with a brush-tip. Of course, at some time, a pioneering chimpanzee must have invented the method for the first time, probably based on other skills already standard in the local repertoire (e.g., the use of brush-tip tools for maximizing the yield of arboreal termites, and the use of strong rods for breaking open bees' nests). In the form it is witnessed in use today, however, the intricate complexity of the skill that is acquired by all the Goualougo chimpanzees bears the hallmark of social learning (Byrne 2007). Acquisition is likely to involve careful examination of the products of skilled tool users—the location and nature of the entry hole, and the tools themselves—and observational learning of the "gist" of the method, such as the way in which one tool is used two-handed with power, another gently with one hand.

Chimpanzees are alone among the apes in regularly making and using several different types of tool in the wild, sometimes even deploying more than one in planned sequence for a single purpose. So their causal understanding must be different in some critical way . . . no? Well, it's not that simple, as Bill McGrew lamented long ago in a paper entitled "Why is ape tool use so confusing?" (McGrew 1989). The biggest problem is that in captivity the other great ape species show similar abilities to chimpanzees in solving tasks that require tool use: gorillas are just as good at getting honey from an "artificial termite mound" by making appropriate probe tools, and orangutans are famous among zookeepers for their technological successes. (An apocryphal story goes the rounds. If a keeper loses his spanner in the chimpanzee enclosure, the chimps smash the glass with it and cause havoc; if he loses it in the gorilla cage, the silverback finds it, examines it carefully, and gives it back to him; but if he loses it in an orangutan cage, it goes missing. Then, two nights later, the orangutans escape by undoing all the bolts of the cage ...) In the laboratory, the "trap tube" task is often seen as a test of the understanding of causality in tool use: invented by Elisabetta Visalberghi, the apparatus consists of a horizontal see-through tube with a peanut halfway along, but on one side there is a well into which the peanut can fall and be trapped. Some sort of probing stick is provided. The trick is always to probe from the side where the well is, so that the peanut can't fall into it. Capuchin monkeys, the species showing most tool use in captivity and recently found to use stones as hammers to crack nuts at a few sites in the wild (Fragaszy et al. 2004), do very poorly at the task. Most never learn to do better than chance, and the few who do turn out to use special-purpose strategies that don't depend on understanding about things falling (Visalberghi and Limongelli 1994). Revealingly, when the tube is rotated so the well points up in the air and can't possibly work as a trap, they avoid it just the same. Chimpanzees do only marginally better (Limongelli et al. 1995). Yet when the tool is removed from the equation, by allowing the chimpanzee to move the food along the tube with finger-holes, chimpanzees suddenly show that they understand the danger of falling down into wells perfectly well, and aren't fazed by upright-pointing wells in the least (Seed et al. 2009). Since this modification has yet to be tried with most animals, we still don't know whether chimpanzee causal cognition is in any way special compared to that of other great apes; and it would be premature to claim that this modification has now given us a pure test of causal understanding.

Ironically, given the dominant role of tool use and tool making in anthropological reconstructions of human origins, it seems we cannot yet conclude much about insight from studies of animals using tools. Even when tools are made, and even when learning to use them involves social learning, the individual may have no insight into how tool use works.

Categories of danger and risk

Almost every animal has ways of avoiding predation by others, but some ways impress us as being smarter than others. For social species, having a system of signals to give the alarm just at the moments when fear-and-flight is appropriate is a great deal more efficient than constantly living in fear and avoiding every location where a predator is likely to be met, and many species of birds and mammals have evolved alarm vocalizations. Not all dangers need the same response, and if the alarm call can prepare hearers to take just the right kind of evasive action it will function better. As we saw in Chapter 3, some species have evolved predator-specific alarms, most famously the vervet monkey. Whereas the degree of response varies with the intensity or repetition of playbacks, the type of reaction is specific to the call type. So, a "leopard alarm" causes hearers to run up trees to the tips of high branches where a leopard can't reach them; an "eagle alarm" causes almost the opposite reaction, with vervets plummeting down from treetops into dense cover. This makes perfect sense, and it is clear that vervets gain useful information from hearing alarm calls. But exactly what information do hearers glean from a call? It must be more than "danger, better watch out," because monkeys react in a way appropriate to the specific class of predator. But what does a vervet hearing an eagle alarm actually think? Is it something we might gloss as an instruction, "dive for thick cover, for some reason"? Or does it refer to the direction from which danger comes, "major danger in the air"? Or maybe it refers to the general class of predator, "eagle"—or more specifically, "martial eagle *Polemaetus bellicosus*"? Or more specific still, "Alex, the male of the pair of martial eagles that nest in the tall acacia by our sleeping site"? The monkeys' reactions can't tell us.

Humans, of course, categorize entities at many different levels: we even have words to describe each level, as I've been using to speculate about monkey classification of their natural predators. Nevertheless, some categories are apparently more natural to us than others, in the sense that those categories are more easily learnt, children acquire them earlier, adults can answer questions about them more quickly, and all languages have simple terms for them. "Natural categories" include drum, but not snare drum or musical instrument; bird, but not song thrush or passerine; car, but not hatchback or vehicle (Rosch et al. 1976). However, it's entirely possible that this privileged, natural level is itself a result of experience, rather than a cause of how the world is perceived. Most people in Western environments have no real need to categorize below the level of "bird," but fanatic birders may find it much more natural to categorize at a lower level—perceiving a bird directly as some kind of eagle, owl, gull, sandpiper, and so on.

As far as I know, a more precise level of natural category has not been proven to exist in birders or any other unusual group of linguistic users, but something like it has been found in the African elephant. Many animals respond to the species "human" in a distinctive way, according to their past experience of humans. At Amboseli National Park, Kenya, elephants meet four different "tribes" of humans: elephant researchers, who travel in Land Rovers and may hang around near them for long periods, but are harmless; tourists, who travel in zebra-striped minibuses, make a lot of dust but don't stay around long, and are harmless; Kamba people, from nearby agricultural villages, who travel on foot and are mostly harmless; and Maasai, who also travel on foot but can be very dangerous. The Maasai are not a hunting people, they are pastoralists who get all they need from their cattle herds; but young men show off their bravery by deliberately spearing lions, and sometimes elephants, simply because it is dangerous. (It is illegal within the National Park, but sometimes still happens just outside.) Over the years, researchers noticed an odd thing: elephants often tend to attack domestic cattle, though these animals pose no possible threat to them. Why? Lucy Bates and I attempted to find out, working on the hunch that it was because the cattle in the park were associated 1:1 with the Maasai, so perhaps the elephants distinguished among human tribes, like we do. We took other common associates of the Maasai lifestyle, visual and olfactory, and presented them (in the absence of any real Maasai people) to groups of elephants, comparing their reactions to the equivalent stimuli from the Kamba people (Bates et al. 2007).

The first comparison was olfactory: we asked local Maasai and Kamba people to wear a garment for a week, and then give it to us, in exchange for a new one. (This was a popular offer: Maasai warriors immediately reached for their cellphones to get our numbers, so we couldn't renege on such a brilliant deal.) Then we set out one of the garments on a low bush, so that a moving elephant group would soon have a chance to pick up its scent: elephants rely heavily on their advanced sense of smell, and in these experiments the elephants seldom actually saw the garment. Their reactions were strikingly different, according to the tribe of the original wearer. With a Kamba-worn cloth, they stopped and sniffed the breeze with trunk up, then continued feeding; but with a Maasai-worn cloth, the group bunched up, trunks uplifted, and pointing in the direction of the scent, milled around, then fled, typically for over a kilometer and often breaking into a run. Subsequently, Karen McComb and colleagues have shown that the elephants also distinguish these same human groups by the sound of their speech, reacting much more strongly to overhearing words spoken in the Ma language of the Maasai (McComb et al. 2014). Intriguingly, when we instead used a visual correlate of Maasai lifestyle—the distinctive red color of their garments—the elephants' reactions were subtly different (Bates et al. 2007). This time, we took clean cloths and placed them so the elephants would sometimes notice them. Although they distinguished Maasai red from innocent white cloths just as clearly, this time they showed approach, aggressive display, and sometimes violence toward the red cloths. For an elephant, olfaction is probably the dominant sense, so it is likely that the absence of any Maasai scent on these cloths convinced them that it was safe to approach: then the elephants showed aggression, giving an idea of their emotional attitude. Incidents in which elephants are speared by Maasai are infrequent, and many of the elephant families in our experiments had experienced none in the 35 years over which they have been closely studied; yet their reactions to the experimental presentations were equally as strong as families who had direct experiences of death or injury from Maasai spears. Thus, the knowledge that one subcategory of the human species poses a very specific risk is likely to have been transmitted socially, rather than relying on individual experience.

Curiosity

If animals only behaved according to basic principles of survival and reproduction, their lives would be entirely filled with the search for key resources: finding food, drink, and mating partners; avoiding undue risks, even when asleep; building useful relationships; rearing offspring—and all the other utilitarian and essential functions biologists study. But sometimes animals also do something else: they explore objects they haven't seen before, and play around with all sorts of apparently useless things. It's tempting to think that, just like us, non-human animals—or at least some of them—show interest in the world "for its own sake." Humans are quite proud of their curiosity: NASA calls its immensely sophisticated Mars exploration vehicle "Curiosity Rover." Should we accept that animals can also show a sort of scientific motivation, a simple curiosity about how the world is? Once you think of animal behavior in information-processing terms, the need for something like curiosity becomes obvious. The point is that, barring animals with the very simplest of lives (limpets?), information is power. Information gathering is worth doing, even if there are no obvious payoffs at the time, as long as getting the information is not unduly costly or risky. Storing information in memory is cheap, and you never know when a little knowledge may come in handy.

Curiosity isn't necessarily a good thing for all animals, though. Think of the old saying: Curiosity killed the cat. Or what happened to Pandora, when her curiosity got the better of her and she opened the box. Investigating things you don't know about, places you don't need to go, individuals you don't need to meet, may have significant costs (think of traps, leopard lairs, and diseases, respectively). For genetical selection to favor curiosity, biological function must trade off against costs: and costs will depend on an animal's ecology. The white rat, that favorite animal of behaviorist studies, is a domestic version of *Rattus norvegicus*, a species that has colonized the globe from obscure origins in Central Asia by adapting and exploiting human ways: a superb generalist. Generalists need to respond rapidly to changing environments, so it pays to explore the world and build up a mental model of what is where and how to get there. In animal learning terms, getting extra information is "rewarding" for rats: they will work for it. Monkeys too, as demonstrated by some original experiments in which monkeys proved willing to work in order to open a blind—which gave them nothing more than a view of a picture (Humphrey 1972). And not just any picture: every monkey tested reliably preferred pictures of other animals, monkeys, or people to those of flowers, food, or an abstract painting (a Mondrian). The monkeys in question were *Macaca mulatta*, the common monkey of northern India: another generalist, well able to colonize cities as well as jungle. Species with very specific niches may be rather more risk-averse and more discerning in what they seek out or work for, since they'd have less to gain by acquiring general knowledge.

Although noticing and remembering factual information—of the who–what–where–when variety—is likely to pay off for many species, it would be even more significant if, like human scientists, some animal species were able to add a how or why. With that level of understanding, it would be possible to compute mentally, on the basis of known facts, whether or not something was likely to happen. We do this all the time, and when we detect a mismatch with what we perceive, we become understandably curious: we use curiosity to better understand the world. This means that some kinds of curiosity have the potential to tell us about the ways in which animals understand their world: cases in which nothing abnormal is present, superficially; but the configuration is improbable and surprising, to those who have a causal understanding of objects or a mental-state understanding of individuals. Consider our own thought processes when we ask: "Why is that thing lying on the ground, just here?" or "How was that thing made, and by whom?" Precisely what information can be extracted from any given situation depends on how that situation is perceived. With more advanced perceptual and brain processes, there is more to discover; with more advanced motor abilities of brain and effectors, more can be done. Inevitably those species with limited perception and restricted ability to affect the environment are not going to show much sign of curiosity; so what animals are curious about, and how long their curiosity lasts, may be revealing of their information-processing abilities.

The seminal work on curiosity was by Steve Glickman and Richard Sroges in the 1960s. They presented over two hundred individual zoo animals with novel objects, identical in form but scaled for body size, and measured how long the animal stayed interested and how many different ways they manipulated them (Glickman and Sroges 1964). The objects were simple ones: two wooden blocks, rubber tubing and wooden dowelling, steel chains, and a ball made of crumpled paper. Primates and carnivores showed much greater interest than rodents, marsupials, or edentates (sloths, armadillos, and anteaters); reptiles showed least of all. Over successive minutes of the test sessions, the level of interest naturally declined, but primates' level of interest held up longer than that of carnivores. Among the primates, the terrestrial Old World monkeys, animals like baboons and macaques, showed greater interest in the objects than other monkeys, manipulating and grasping them under close visual inspection, as well as simply looking at and chewing upon them. (Testing adult great apes was not possible, for safety reasons.) There were no general sex differences in curiosity, but younger animals showed greater interest than adults.

There have been few similar studies since: sadly, as this simple approach shows great promise for making systematic comparison among many species. It would be fascinating to know whether differences in ecology (extractive foraging, diet breadth, etc.), social system, or brain size best predicted levels of curiosity among closely related species. Yet Glickman and Sroges were essentially given the run of Lincoln Park Zoo, Chicago, and were able to separate individuals at will and test when the animals were off view in the evenings; and zoo animals in the 1960s usually lived in bleak cages, so were starved of interesting things to examine. Improved welfare for zoo animals, with restrictions on separation of individuals and high levels of enrichment—making novel objects like chains and blocks pretty uninteresting to modern zoo animals—combined with a health-and-safety culture, would make the experiment hard even to replicate now.

Summary

The abilities shown in dealing with the physical environment are evidently important as a basis for developing insight into physical causes. These capacities include: remembering what was noticed where and when; being able sometimes to predict changes to those objects when away from observation, such as decay of perishable foods, ripening of fruit after warm weather, or food pilfering by competitors that saw it cached; planning efficient routes to remembered locations even if that particular route has not been traversed before; distinguishing among sources of danger and risk in a way that reflects the evidence, whatever the level of abstraction needed; and collecting knowledge that might pay off in the future, by means of natural curiosity and a tendency to explore. This list includes some remarkable animal abilities, only discerned in recent years, which must rely on representational understanding, especially where new information must have been inferred. But we have seen little sign that special abilities are associated with just those species showing insight in other spheres. Rather, most of the capacities seem general across many species—at least, those species whose ecology is best exploited by their use. Often the best evidence comes from the well-studied primates, but there is little reason to suppose that primates are in any way unique in these capacities, and certainly the great apes do not stand out on current evidence. The exception is in the case of tool use, where the chimpanzee is unique in its routine use of tools for many purposes in all groups studied in the wild, and the other great apes show similarly remarkable capacities for tool use in captivity. Yet even here, laboratory studies of causal cognition have failed to identify superior understanding of everyday physics by great apes. I suggest, therefore, that understanding of the physical world does not differ greatly among a wide range of animal species. Indeed, the physical world is likely to pose similar problems, solved by similar solutions, to a wide range of animal species. Furthermore, the distribution of insight—that is, specifically in the great apes and a few other large-brained species—that we have noted in this book up to now is not explained by variations in physical understanding. Something else is needed, building upon the impressive representational understanding of physical objects and what happens to them, general among many species—something that singles out those few whose insight applies more generally. For that, we must turn to Chapter 11, and return to the consequences of social living: in particular, the opportunities for exploiting the physical world that are offered by living with others who already know how.

Chapter 11

Learning new complex skills

Behavior parsing and the origin of insight

When we approach any of a wide range of problems, from car maintenance to public speaking, we do so with a pre-existing repertoire of actions ready to deploy. Some of these actions are no doubt innate, and many others are built up during our lifetime by trial-and-error exploration of previous similar situations; but a significant part of our repertoire of actions is learned by noticing and using other people's behavior. With the ever-increasing pace of change of twenty-first-century life, we rely more and more on the last of these methods for learning new or smarter ways of doing things, often things involving technology. And despite the complaints of the older generation, it's really remarkably easy—you just watch someone using the new gear, pick up a few tips, and you're (usually) in business. It comes as rather a shock to realize that most of the animal kingdom simply can't do that. Nor does it help prepare us for that shock that in everyday talk, even in wildlife documentaries on the media, people routinely describe young animals as imitating their parents or other knowledgeable adults in order to learn survival skills. Yet many psychologists would doubt that any animal can truly imitate. What's the problem? What are these animals doing instead that is glossed as imitation? If learning by imitation is so hard, how do we do it so effortlessly? And if imitation is so rare, does its pattern of occurrence correspond with the occurrence of insight? Is imitation the "missing link" that we alluded to in Chapter 10, between representing the physical attributes of the world and thinking about its social aspects?

There are two important points to make that help us thread a path through this area: first, imitation is much more than just socially influenced learning, something which is seen in a great range of animal species and about which there is no controversy; and second, despite the fact that humans can readily learn by imitation, we should remember that our learning processes still rely heavily on the same mechanisms as those of most other animals, it's not one or the other. Starting with a brief recap of social learning theory, based on the fuller account in Chapter 7, let's try to home in on the very special ability of imitative learning.

Managing without imitation

In all animals, human or otherwise, individual exploration is responsible for a great deal of effective skill acquisition. This has the great advantage that the result is reliable (compared with social learning from a model who may be performing suboptimally), and the behavior acquired will automatically be tailored to the individual concerned (compared with imitative copying of someone who is perhaps bigger and stronger than you, and can do things you simply can't).

In social species, individual learning may be guided or encouraged in the right direction merely by the presence of others, in several ways. Simply by accompanying those who already know how to do well in the natural environment, a social animal will be taken to places where success is definitely possible, and away from places where life is particularly dangerous and exploration highly risky. Moreover, this will afford them the chance to explore environmental debris resulting from others' behavior, such as feeding remains or discarded tools, and what they learn about these objects may result in more rapid acquisition of the necessary skills for using them. Whether these mechanisms are called social learning or "socially guided learning" is a moot point, but in any case they are benefits of social living that come free.

Two other processes require some input from the learner, but their benefits are so clear and the cognitive mechanisms required are so simple that it is likely that they are very widespread among social animals. The first of these general-purpose social learning devices involves restricting an individual's exploration to suitable things. Suppose an individual sees another of the same species engaged in some activity, and as a result their attention is drawn specifically to the place where the other is busy (hence one name for the effect, "local enhancement"), or drawn to a range of associated environmental aspects, including the objects that are being processed and any tools involved (hence the other name for it, "stimulus enhancement"). Once the observer's exploration has been directed to the objects that a (skilled) conspecific was using in a specific place, its own efforts are much more likely to lead to discovering the methods that should be employed than if it had to start cold. The second general-purpose device is a matter of encouraging the most suitable kind of exploration. Many animals have quite extensive repertoires of actions: if inappropriate behavior is tried out, even with the right things in the right place, success may be slow. In "response facilitation," if a conspecific is seen to be using an action that is familiar to the observer (i.e., an action in its own repertoire), then it is more likely to try out that action than others in its repertoire when it gets the chance to explore the situation. Both stimulus enhancement and response facilitation may be two aspects of the same process: if brain records are automatically "primed" or activated by what is seen in a social situation, and an individual has brain records for places and objects, stimulus enhancement will result; if it has brain records for action patterns in its repertoire, response facilitation will be produced.

The combination of these simple effects is a powerful one, and the results can look very like imitation. Imagine an experiment in which an ape, taken from a population which has no access to nuts, is introduced to another in which cracking hard nuts with stones is regularly seen. Initially, the newcomer knows nothing of nuts and gets somewhat hungry. Then, as it becomes integrated into the community, it is permitted to accompany groups that travel to their habitual nut-cracking areas, where it can play about with the loose stones and broken nut remains lying about. It watches other apes picking up stones and bringing them downward forcefully onto hard objects that do not look very edible, and it notices that they are eating with enthusiasm. As a result of its own exploration of the area, guided by feedback from trial-and-error learning, including manual exploration directed specifically to the nuts and rocks it has seen others pick up (by stimulus enhancement) and prominently featuring downward smashing actions (by response facilitation), it discovers how to get food from the hard objects it might otherwise have ignored forever. Media commentary might describe the ape as imitating the skills of its new community members, but my account has studiously avoided recourse to imitation as an explanation. This particular experiment has not, to my knowledge, been done, but observations of the general kind have been made many times with many species, and are often called imitation in popular accounts.

How could we recognize real imitation, then, against a background of social learning that is made so efficient by the simple mechanisms of priming? Most researchers use Thorndike's classic definition of imitation "learning to do an act from seeing it done" (Thorndike 1898); so the critical question becomes, does the learning result directly from seeing the act done (imitation), or does seeing the act done merely affect the process of exploration, indirectly causing learning to be more efficient? From the descriptive account in the last paragraph, it would be hard to tell for sure: maybe the ape did actually learn that smashing down rocks was the way to get edible matter, when it saw it done. It might have remembered that information, even if it had no opportunity to put it to use at the time, and used the trick next week when it got the chance. Or maybe not: most laboratory experiments that claim to have shown imitative learning in animals test them soon after they've had the chance to observe a successful model, and so are ambiguous. They do not rule out the possibility that priming caused those same actions to be tried out sooner than they might otherwise have been done, and since they work, and the animal gets the food reward, it learns to use them again as a result of its trial and error.

Different kinds of imitation: Contextual and production

There are a few laboratory experiments, however, which do rule out explanation by response facilitation alone. In one, budgerigars were shown one of two equivalent ways of getting to a food reward, used by a pre-trained budgerigar in each case, and then made to wait for some time after this observation before they were tested (Heyes and Saggerson 2002). In another, quail were shown one of two actions used by pre-trained quail successfully to get food, but were then distracted by being made to perform other activities during a wait before testing (Zentall and Akins 1996). In both cases, the birds showed they had learned which act to do, from seeing it done. Thus, at least two species of bird (and, we'll assume, humans) can go beyond priming: they can imitate. But what kind of imitation is this? There is more than one!

In those quail and budgerigar experiments, the new information learnt was solely about when and where to use an already-known action, rather than about learning a new action. I've called this *contextual imitation*, to distinguish it from the more everyday sense of imitation in which a new pattern of behavior is acquired from seeing it done, *production imitation* (Byrne 2002b). If we learn by imitating someone else to turn a screw clockwise rather than anticlockwise, to lift a catch to open a gate, or to press the end of a new ballpoint to reveal the tip, then we have learnt by contextual imitation, because the actions of turning a screw, lifting a catch, or pressing a pen-top were not new to us when we saw them done. The contextual/production distinction was first used by Vincent Janik and Peter Slater to distinguish different types of learning (Janik and Slater 1997): in contextual, or usage, learning, we learn when to apply an action we already know; whereas in production learning our repertoire of actions is increased.

Contextual imitation adds power to social learning, and may be responsible for perpetuating many of the traditions of behavior observed in animals, such as the ways that birds gain the food we put out for them in gardens, as described in Chapter 7. But it has its limitations, because the imitated actions must already be part of the individual's repertoire. New actions can always be learnt, of course, by trial-and-error learning, which—as we've seen—can be made much more powerful in social contexts by stimulus enhancement and response facilitation. It is only in cases where an action is not already in the natural repertoire of the species, and *not at all likely to be discovered* by trial-and-error exploration, that production imitation becomes really useful. Most animals manage perfectly well without it.

Where might we find evidence for production imitation in animals, if at all? We have already met one likely candidate: the complex food-extraction and food-processing skills of the great apes, including the tool-making and tool-using abilities of chimpanzees and some orangutans. In Chapter 9, I argued that it was the intricate complexity of the ape skills, rather than the use of a tool per se, that made them so special. In Chapter 10, we saw that there was no simple explanation in terms of their physical cognition that could explain why apes usually can, and monkeys usually can't, develop these elaborate tool-using skills. Here, I suggest that the great apes—but perhaps few other species—can learn new skills by production imitation, and that this is what has led us to find feeding skills of such remarkable complexity in these species alone.

To make this claim at all convincing, it's important to demystify the process of production imitation itself, by taking a cognitive approach and examining just what is necessary to do it. Two components are required: first, the fluid performance of a skilled action must be analyzed into the component units from which it was built; then, the organization of these units into a planned action must be detected. As humans, we find it easy and natural to "see" others' behavior in this way, but neither of the processes is a trivial one.

Segmenting the stream of action

When we watch skilled performers of practiced routines, like a mechanic fixing a car or a cook making a meal, or listen to a native speaker chatting away, the behavior that we perceive does not come with ready-made gaps that correspond to logically distinct elements. This has been classically noted to apply to speech, where a sound-gap is more likely to be part of a plosive consonant than to signal a new word, but in fact the point applies to all skilled motor action. The physical stimulus that confronts us is smooth and fluid, like a river, not segmented into chunks, like a freight train. The remarkable thing is that we are nevertheless able to segment that smooth and apparently unbroken flow of action into sensible chunks.

What is meant by "sensible," here? To answer that, we have to think more carefully about this idea of a chunk of action that I've casually introduced. I propose that people are able to "see" (pick out within a stream of action) any element which is already present in their personal repertoire (Byrne 2003b). If you already know how to undo a nut, you notice the mechanic undoing one; if you don't, you may instead notice a repeated turning movement of a wrench. (Likewise in speech, if you already know the name "Misato," you notice it in a conversation; if you don't, you may only hear "Miss" and struggle to make sense of the "ato.") The "size" of each element is therefore irrelevant. For different observers, or at different times in the life of a single observer, one particular movement of a single finger or an elaborate sequence of bimanual movements might both properly be seen as single elements.

When we watch a relatively unfamiliar process being performed, the analytical level at which we notice elements will be low, perhaps that of finger movements; whereas when we watch a slight variant of an already familiar activity, the basic elements that we notice might themselves be high-level, complex processes. Most commonly, perhaps, the level at which observed behavior matches parts of our existing repertoire would be neither of these extremes, but rather consist of simple and highly practiced movements that produce visible effects on environmental objects: that is, simple, goal-directed movements (Zacks et al. 2001). Such elements may be particularly easy to delimit because they are marked by a characteristic pattern of acceleration and deceleration, just like the cadence of syllables in a sentence (Zacks 2004; Zacks et al. 2009). Consistent with this idea, people are able to pick out the basic structure of action, even when the stimulus is experimentally reduced to the point when the activity is unrecognizable. Dare Baldwin filmed people carrying out everyday activities, like whisking cake materials together or washing up, with fluorescent spots on key joints and fingers; then, the films were processed until only the movement of the glowing spots remained. When these dot-dances were shown to subjects they could not guess what activity was being shown, but they could parse the activity into its logical phases (Baldwin et al. 2008; Loucks and Baldwin 2009).

The idea that each element "noticed" must already be within the repertoire of the observer has an automatic and very convenient consequence. Any such element can immediately be used as a building block in effective motor planning, just because it is already in the observer's repertoire—whether it is a finger movement or an elaborate but familiar routine. So, segmentation on this principle automatically provides a stream of elements that can be replicated by the observer, and forms the first component of my cognitive model of imitation (Byrne 2003b). Segmentation of action into units that "make sense" because they are also in the observer's repertoire of action is, however, a process that may be useful for other things than imitation. Indeed, as we'll see next, there is evidence that this ability has evolved in animals that cannot imitate; evidently it must have evolved because of a different adaptive consequence. It is therefore likely that action segmentation is the most primitive part of the cognitive system for imitation, one inherited from ancestors who could not themselves use it for imitation, but forming the key basis for imitation to operate upon. A digression into recent neuropsychological studies of monkeys shows how this might have happened.

Non-human primates have been shown able to pick out, in the behavior of others they observe, actions that are already in their own repertoire. A system of single neurons has been identified in the premotor cortex of rhesus monkeys (Gallese et al. 1996; Rizzolatti et al. 1996, 2002), each of which responds to a simple manual action, and responds equally whether the monkey makes the action or sees another do it. The cardinal properties of these "mirror neurons" are that (1) they detect goal-directed movements that are in the observing monkey's own repertoire, and (2) they generalize over whether the movement is performed by the monkey itself or by another agent.

These units have sometimes been described as "monkey see, monkey do" cells, but rhesus monkeys are not among the animal species found capable of imitation, despite their long history of study in psychologists' laboratories; in fact no monkey species has been convincingly shown to learn new skills by imitation. Monkeys also lack the elaborate food-processing skills of great apes (though see Chapter 7 for capuchins that use stone tools), so there is no strong reason to attribute the mirror neuron system to part of an imitation mechanism. Giacomo Rizzolatti and his collaborators have a quite different suggestion: that the system functions in revealing the demeanor and likely next actions of conspecifics, by reference to those actions the observing monkey might itself have done (Rizzolatti et al. 2002). That is something which monkeys are exquisitely good at, and on which they rely every minute of the day for avoiding the pitfalls of their socially complex groups. While the mirror neurons that allow detection of a conspecific's demeanor are no doubt likely to be innate, there is no reason why experience might not "train" mirror neurons to have more general properties: Celia Heyes has developed a model in which the connections between actions-as-seen and actions-as-done could be acquired by experience (Heyes and Ray 2000).

Despite the likely origin in social perception, there is a sense in which the monkey-see-monkey-do label is accurate. The mirror neuron system may be the neural basis of response facilitation, tuning a monkey's exploration toward the actions it sees others using around it. And response facilitation may be the fundamental first-pass process that allows both forms of imitation to operate.

Action-level imitation

Response facilitation, operating by means of mirror neurons, responds to precisely those movement patterns that correspond to actions latent in the viewer's repertoire. Response facilitation therefore automatically analyses a continuous flow of observed movements into a string of recognized, familiar actions that are each present in the observer's repertoire. Once the fluid stream of action has been segmented in this way, then action-level imitation becomes feasible. All that is required is some form of memory in which to hold the assembled motor program. Heyes' model, mentioned previously, included the suggestion that sequential links between the actions in a regularly observed sequence would automatically be learnt, allowing (action-level) copying of sequences of acts from seeing them done.

Surprisingly, then, there is only sparse evidence for action-level imitation in non-human animals. The single items for which contextual imitation has been shown in budgerigars and quail—foraging movements typical of the species—might be thought of as the simplest case of action-level imitation, each a string of one. Chimpanzees have been found to copy the order of three actions, even though the sequence was entirely arbitrary and unrelated to success (Whiten 1998). The limitation for most animals may be on working memory capacity, rather than in ability to extract the string of elements to be copied. Humans show greater facility at action-level imitation: consider the graceful routines of modern dance, or the clever movements of a great tennis player. However, the difficulties we have in learning these things entirely by imitation suggests that action-level imitation is hard work: usually our learning is propped up by verbal cues and tricks, even in cases where ultimately seeing is necessary for good copying.

In action-level imitation, a linear sequence of actions is copied without recognition of any higher-order organization that may be present: the organization is "flat." There may in fact be relatively few cases where it is beneficial for animals to copy behavior that is genuinely linear in structure. Building up linear sequences of actions by trial-and-error learning may be slow, but—as noted earlier—it is also reliable, in that only sequences that work well for the learner will be learned. Relying on action-level imitation for this job is less secure.

Program-level imitation

Most human action, and arguably also much of the behavior of non-human great apes, is planned. Planned action has hierarchical, not linear organization: it is composed of embedded routines that deal systematically with sub-goals, and is only linear in the final output sequence. To fix a flat tire, the car must be jacked up, then the wheel replaced, then the car lowered. To jack the car up, the jack must be found, fitted under the right place, then the car jacked high enough. To get the car high enough, the jack handle must be turned and the height change checked: this routine must be repeated until the right height is reached. And so on. An observer will see you doing "one thing after another," but that is not how it was planned, and it would be laborious and inefficient to copy it that way.

Copying the organization of planned behavior is likely to pay handsomely, for an animal that can do it. To distinguish this kind of copying, in which the hierarchical structure of behavior is detected from the observed actions of others, from action-level imitation, I've coined the term *program-level imitation* (Byrne and Russon 1998). But don't let the fancy term fool you: it is what we do all the time. It comes naturally to us, as humans, and "program-level production imitation" is what we normally just call imitation!

The next section of this chapter considers whether, and how, a bottom-up, mechanistic cognitive analysis can explain how behavioral organization can be parsed out from watching someone's behavior, and how thereby that behavior can be copied. The big question is, are we alone in this ability? Or can non-human great apes, for instance, also convince us that they have something of our facility in program-level imitation?

Imitation in great apes

As we have seen, most animals do not have the ability to learn by imitation at all: they learn socially, but they do not imitate. But then, most animals do not learn sufficiently complex patterns of behavior to have much need of imitation. Part of the reason for this is that they do not have anything as deft and flexible as a hand: the motor skills you can acquire are pretty limited if you have only a hoof or a flipper. Even relatively dexterous animals like raccoons, squirrels, or otters can only hold objects by using the two hands together, carrying out symmetrical movements. The situation is more interesting for primates. The primitive five-fingered primate hand (Napier 1961) is highly effective as a manipulator and in many species shows some opposability: monkeys can easily pick up quite small objects in one hand. Nevertheless, it seems that monkey manual skills are in the wild generally limited to performing a range of species-typical actions with deftness and precision, rather than building up more complex organizations of behavior (Christel and Fragaszy 2000).

In great apes, the hand shows a considerably greater range of aptitudes compared even to those of monkeys (Napier 1961). To illustrate these differences and their consequences, I will use the example of everyday food preparation in the mountain gorilla. At Karisoke in Rwanda, a gorilla's four main foods all involve using the two hands in different but complementary roles. This is called *manual role differentiation* (Elliott and Connolly 1984). We humans take it for granted, using manual role differentiation a hundred times a day: opening jars, cutting paper with scissors, tying shoelaces, eating at table, and so on. It comes as something of a shock to realize how few other species share the ability with us, but all the great apes do. The sophistication in manipulation that manual role differentiation allows is further increased by the gorilla's ability to control individual digits of the hand independently. This is called *digit role differentiation* (Byrne et al. 2001a). Items can be held in part of the hand, while other digits can carry out other activities: for instance, part-processed food can be retained in the hand by the lower fingers holding the already-processed items, while a section of the food-processing routine is iteratively repeated to build up a larger handful of food, as the forefinger and thumb collect more food items. Again, we take this ability for granted. If you're picking berries from a bush, or retrieving buttons fallen to the floor, it would seem absurd to pick them up and put them away one at a time; but the dexterity that allows us to perform these simple actions efficiently is a rare thing in nature, essentially restricted to our closest primate relatives, the great apes.

Manual and digit role differentiation abilities allow the mountain gorilla to deal with plants that are physically defended by an array of spines, stings, and hard casings (Byrne 2001). In the process, gorillas display a huge repertoire of functionally distinct actions (i.e., single actions that produce clear changes to the plant substrate; thistle-processing alone gave evidence of 72 distinct actions). Although attention has been drawn away from the chimpanzee's general manual skills by the anthropological emphasis on tool use, when chimpanzee plant-processing has been studied in detail, similar abilities are found to those of gorillas (Corp and Byrne 2002a,b). With animals of such dexterity, manual behavior is sufficiently rich for complex organizations of learnt behavior to be detectable by researchers: and it would certainly pay the apes to be able to learn these elaborate skills by imitation of others.

The evidence that great apes do indeed learn skills by imitation comes from observational data rather than experiment, since no useful experimental test of program-level imitation in animals has yet been devised. Although the evidence is therefore oblique, cumulatively it is fairly strong (Byrne 2002b, 2005). For a start, there is the very fact that young great apes learn complex, hierarchically structured routines of manual behavior (some of them essential to survival in adulthood) in just a few years before their weaning, in contrast to monkeys where there is no evidence of anything comparable. Evidence of complexity is strongest for the mountain gorilla, where five-stage sequential processes have been described (Byrne 1999c; Byrne and Byrne 1993; Byrne et al. 2001b; see Figure 11.1), but also clear in chimpanzees, both in tool-using tasks (Boesch and Boesch 1990; Goodall 1986; Matsuzawa 2001; Matsuzawa and Yamakoshi 1996) and in dealing with complicated plant foods (Corp and Byrne 2002b; Stokes and Byrne 2001). The fact that orangutans sometimes also use tools to deal with complex plant defenses (Fox et al. 1999) suggests that they have similar abilities, and this is confirmed by studies of young orangutans' efforts to deal with the vicious spines of certain palm trees (Russon 1998). Far more studies have been carried out on the foraging behavior of monkeys than that of apes, yet no comparable evidence has come to light.

Additionally, in a detailed analysis of variation in the skills of adult mountain gorillas, it was striking that minor details (grip type, hand preference, precisely which fingers were employed, extent of movement) varied idiosyncratically between

144 | BEHAVIOR PARSING AND THE ORIGIN OF INSIGHT

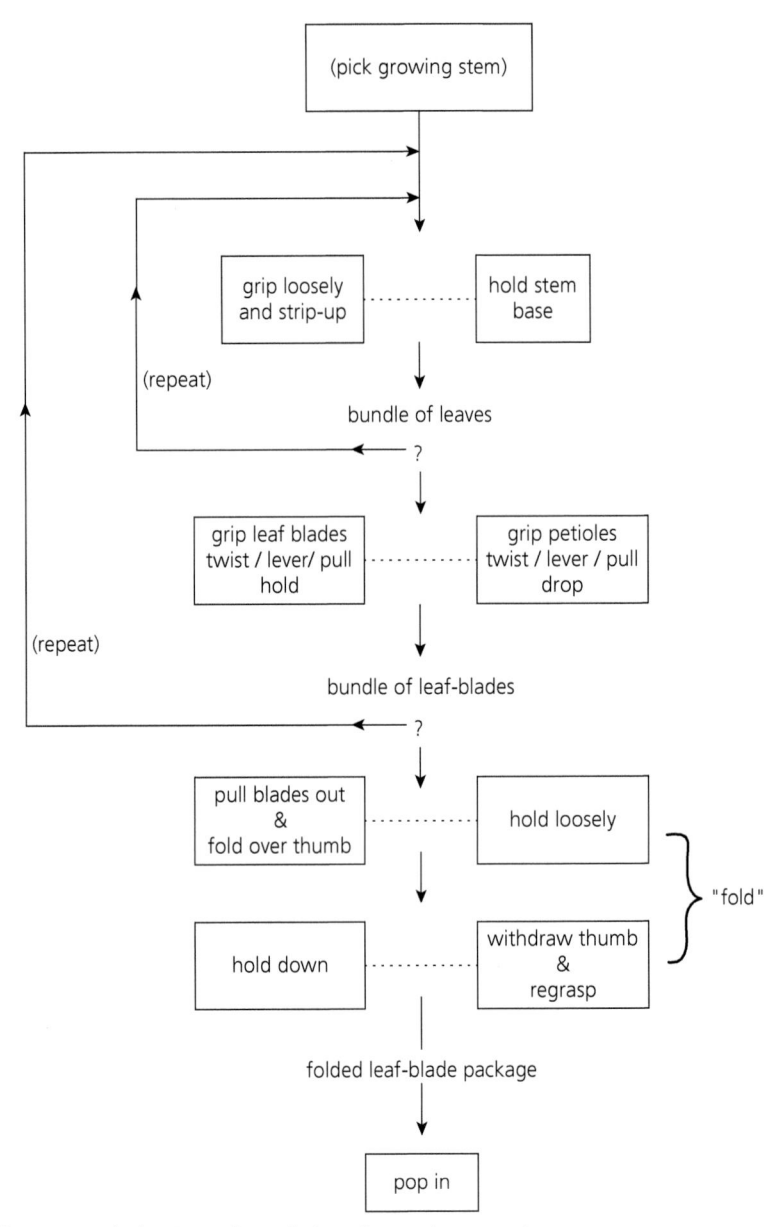

Fig. 11.1 Karisoke, Rwanda, technique for nettle processing.

In this schematic of plant-handling processes used by adult mountain gorillas in Rwanda to eat nettle *Laportea alatipes*, the flow of action proceeds from top to bottom, and optional actions are shown in brackets. Actions that are significantly lateralized to one hand or the other in most individuals are shown on the left or right of the diagram. Some individuals follow this exact organization, others the mirror reverse, with significantly more than half the local population individuals, even between mother and offspring, whereas the overall "program-level" organization of each technique was remarkably standardized in the local population (Byrne and Byrne 1993). The same was found true of chimpanzee feeding techniques (Corp and Byrne 2002b). Such standardization of techniques needs explaining. There are two possibilities: either the affordances of the ape's hands, combined with the physical form of the plant defenses, define a clear gradient of optimization and thus with practice every ape will inevitably acquire the same method (Tomasello and Call 1997); or, observational learning is involved, and key aspects of a skill's behavioral organization are passed on by social learning.

A third line of evidence helps resolve which of these two possibilities is actually the case. It involves the study of gorillas and chimpanzees disabled by crippling snare wounds. Snares are typically set to catch antelopes or wild pigs, but young gorillas or chimpanzees may suffer injury because of their natural curiosity (Stokes et al. 1999). If the standardized pattern of an adult technique is a product of constraints and affordances, then an animal with severely maimed hands would end up with a quite different technique to able-bodied individuals. Yet in both chimpanzees and gorillas, disabled individuals acquire the same organization of behavior as the able-bodied, and instead work around their difficulties by modifying the low-level details of implementation (Byrne and Stokes 2002; Stokes and Byrne 2001; see Figure 11.2). This favors the hypothesis that the standard technique is a socially transmitted pattern.

The plants eaten by mountain gorillas are temperate species, whose distribution is restricted to isolated high mountain areas in central Africa, and all the

---

Fig. 11.1 (Continued)

using the right hand for the deft, delicate operations. Coordinated, asymmetric bimanual actions are shown by a horizontal dotted line: in that case, the two actions must be synchronized to achieve a single result.

Leaves of *Laportea alatipes* are high in protein and low in indigestible fiber, but the plant is protected by painful stinging hairs, especially numerous on leaf-petioles and upper surfaces. The gorillas' processing technique is highly efficient in maximizing intake of leaf-blades while minimizing contact with the stinging hairs by the hands and especially the mouth: it is therefore a valuable part of the mountain gorilla's feeding repertoire. The full technique has six stages. In the first, which is not always performed, the action of picking the whole stem is one common to processing most other plants. The second action, stripping off the leaves from a stem held with the other hand, is also seen when these gorillas strip off (unprotected) leaves from vines, and when chimpanzees sweep driver ants off a stick tool held in the other hand: it is presumably part of the natural repertoire shared by all African great apes. The remaining four actions are probably unique to nettle eating, which is only done by the higher-altitude groups of the Virunga population of mountain gorillas: the problem these four actions solve is only met in nettle eating, and *Laportea alatipes* is a temperate plant, absent from most of the gorilla range.

146 | BEHAVIOR PARSING AND THE ORIGIN OF INSIGHT

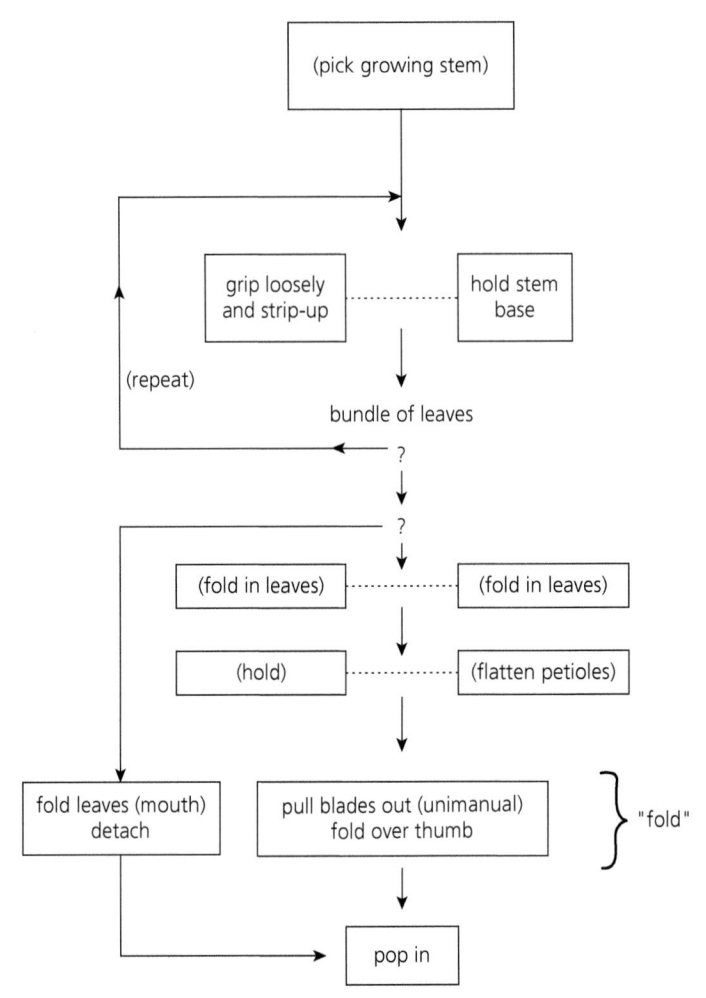

```
                        ┌─────────────────┐
                        │(pick growing stem)│
                        └─────────────────┘
                                 │
          ┌──────────────────────┤
          │                      ▼
          │   ┌──────────────┐  ┌──────────┐
          │   │ grip loosely │··│hold stem │
          │   │ and strip-up │  │  base    │
          │   └──────────────┘  └──────────┘
 (repeat) │          │
          │          ▼
          │    bundle of leaves
          │          ?
          └──────────┤
                     ▼
                     ?
                     │
         ┌───────────┼───────────────┐
         │           │               │
    ┌─────────────┐  │         ┌─────────────┐
    │(fold in leaves)│··········│(fold in leaves)│
    └─────────────┘  │         └─────────────┘
                     │               │
       ┌──────────┐  │         ┌──────────────┐
       │  (hold)  │············│(flatten petioles)│
       └──────────┘            └──────────────┘
         │                            │
         ▼                            ▼
  ┌──────────────────┐    ┌──────────────────────────┐   ⎫
  │fold leaves (mouth)│    │pull blades out (unimanual)│   ⎬ "fold"
  │      detach      │    │    fold over thumb       │   ⎭
  └──────────────────┘    └──────────────────────────┘
         │                            │
         └──────────────┬─────────────┘
                        ▼
                   ┌─────────┐
                   │ pop in  │
                   └─────────┘
```

Fig. 11.2 Modified technique used by hand-injured Pandora.

(Notation as in Figure 11.1.) A number of individuals in this gorilla population suffer from snare injuries to the hands or feet: snares are set illegally in the National Park to catch small antelopes for food, but curious young gorillas investigate unusual objects and are all too often killed or maimed as a result. Pandora's injuries were particularly severe, with only part of the palm and the lowest segment of the thumb remaining on her right hand, and two fingers non-functional on her left. Yet she managed to live a normal life and gave birth to several healthy youngsters who remain part of the Karisoke population today. How did she cope with feeding on a difficult-to-eat plant like *Laportea* nettles? As can be seen from the diagram, the answer is, essentially the same way as other adult gorillas, but modified in several small ways to achieve low-level, local results in a slightly different manner: for instance, using the mouth as well as both hands to achieve the folding operation which an able-bodied gorilla does with its hands in every case. Also note that, despite the impressive level of processing Pandora achieved, the step of detaching the leaf-petioles (stalks covered in stings) is omitted, so probably she suffered somewhat more while eating nettles than an able-bodied gorilla would, even though the petioles were folded into a final package before popping it into the mouth.

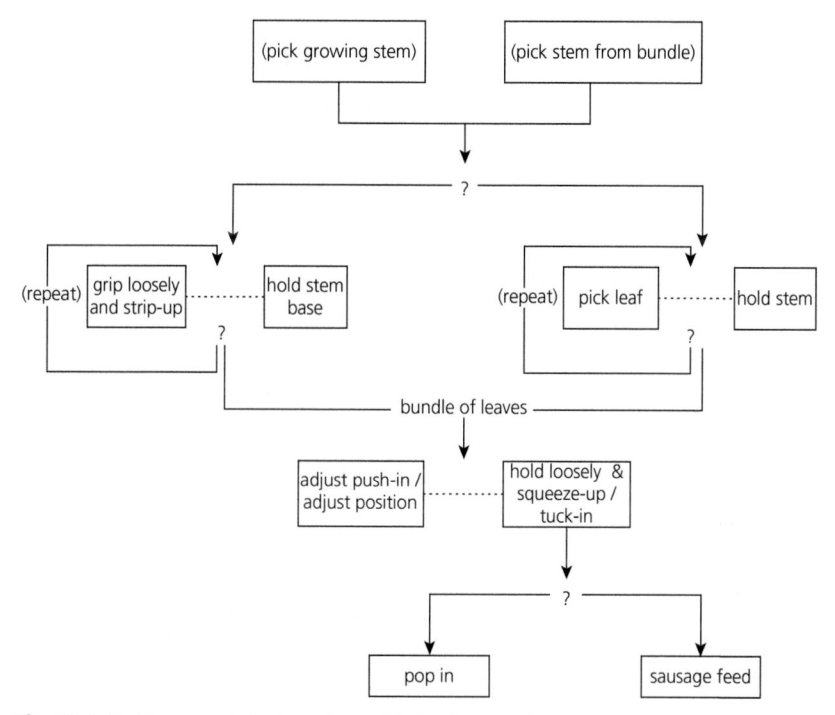

```
                    ┌─────────────────────┐         ┌─────────────────────────┐
                    │ (pick growing stem) │         │ (pick stem from bundle) │
                    └──────────┬──────────┘         └────────────┬────────────┘
                               └──────────────┬──────────────────┘
                                          ── ? ──
                    ┌──────────────┴──────────────────────────────┴──────────────┐
                    ▼                                                            ▼
        ┌──────────┬─────────────────┐  ┌──────────┐       ┌──────────┬───────────┐  ┌───────────┐
        │ (repeat) │ grip loosely    │  │hold stem │       │ (repeat) │ pick leaf │  │ hold stem │
        │          │ and strip-up    │··│  base    │       │          │           │··│           │
        └─────┬────┴─────────────────┘  └──────────┘       └─────┬────┴───────────┘  └───────────┘
              │           ?                                      │       ?
              └───────────────────────┬──── bundle of leaves ────┴──────────┐
                                      ▼                                     ▼
                          ┌─────────────────────┐         ┌──────────────────────┐
                          │ adjust push-in /    │         │ hold loosely &       │
                          │ adjust position     │·········│ squeeze-up /         │
                          │                     │         │ tuck-in              │
                          └─────────────────────┘         └──────────┬───────────┘
                                                                ── ? ──
                                                    ┌───────────────┴───────────────┐
                                                    ▼                               ▼
                                              ┌──────────┐                  ┌───────────────┐
                                              │  pop in  │                  │ sausage feed  │
                                              └──────────┘                  └───────────────┘
```

Fig. 11.3 Port Lympne technique for nettle-leaf processing.

(Notation as in Figure 11.1.) Several groups of western gorillas have been kept at two private zoos in Kent for many years, Howlett's and Port Lympne, where they breed successfully. Some years ago, the keepers began offering some gorillas at Howlett's fresh stems of the European nettle *Urtica dioica*, a plant which is structurally very like *Laportea alatipes* and similarly nutritious, and they evidently enjoyed eating the plants. Later, some of these gorillas moved to Port Lympne where the nettles grow naturally, and since then most of the group there have come to eat nettles regularly. Presumably because the *Urtica* stem is softer than that of *Laportea*, they also sometimes consume both leaves and stem together. Mostly, however, they prefer to eat the leaf-blades, just as do the wild mountain gorillas at Karisoke, but they have developed a very different method as their standard, as shown here. Instead of the elaborate folding process, which envelops the stinging edges and upper leaf surfaces in a single leaf, with underside outwards, they use the action of squeezing the leaves together and tucking in stray bits with the other hand. This approach is sometimes seen in young mountain gorillas, but they go on to learn how to fold. In addition, the Port Lympne gorillas sometimes strip stems of leaves, as at Karisoke, but also use the laborious method of picking single leaves off the stem and accumulating them in the hand. Again this is seen in young mountain gorillas, and perhaps has some advantage in minimizing stinging, although it is clearly less efficient. The difference here may be due to the difference in pain experienced from the stings. Finally, at Port Lympne, leaf-petioles are not detached. Since the petioles are consumed, despite their many painful stings, and the stinging leaf-edges and upper leaf surfaces are not wrapped in a folded leaf, the Port Lympne method is inferior, though it works. Nettles have perhaps not been available for gorillas in Kent for long enough for the local tradition of consuming them to have adjusted to the most efficient method; and the western gorilla is a tropical species, so even the few wild-caught individuals would not have encountered nettles in the wild. At Karisoke, in contrast, nettle eating has been a traditional part of the population's skill repertoire for many generations.

mountain gorillas whose food-processing skills have been studied were part of a single contiguous population in the recent past. However, one captive population of gorillas, originally from quite a different area of Africa and now at the zoos of Howlett's and Port Lympne in Kent, has fortuitously been exposed to nettles as a food source for many years. While these nettles are of a different species to those eaten by mountain gorillas in Rwanda, the technical problems they present for pain-free eating are very similar. If the Karisoke nettle-eating technique is a result of affordances of gorilla hands and plant constraints, we'd expect a similar technique to have developed in Kent. When a study was conducted at Port Lympne zoo, the gorillas were found to have a different technique entirely. It achieved the same result, though rather less efficiently than Karisoke gorillas—as would be expected, when the Kent tradition had only had a decade or so in which to develop (Byrne et al. 2011; see Figure 11.3). Again, social transmission of specific behavioral techniques is favored as the explanation, for both Rwanda and Kent.

Finally, one anecdotal observation supports the case that great apes require imitative learning to acquire certain aspects of their complex feeding techniques. One single adult in the Karisoke gorilla population differed in the technique she used when processing stinging nettles: the female Picasso did not fold bundles of leaves, so was presumably often stung on her lips (Byrne 1999b). Picasso had transferred into the study area from lower altitude, where nettles do not grow. Because adult gorillas feed alone and out of sight of others in dense herbage, mountain gorillas' only opportunity for observational learning of plant processing comes in infancy. It seems most likely that a lack of opportunity to observe any skilled model accounts for Picasso's incomplete technique; and intriguingly her juvenile was the only other gorilla in the study population to lack that particular element of the skill.

Parsing hierarchical structures of behavior

The conclusion that great apes learn new manual skills by program-level imitation, and that that is how their impressive feeding techniques are acquired, remains controversial (Tennie et al. 2009). And indeed, when psychologists talk about learning by imitation, it can sound like pretty complicated and clever stuff. For instance, "the child must imaginatively place herself in the circumstances of the adult and determine what is the purpose of the behavior and how one goes about accomplishing that purpose" (Tomasello et al. 1993). That sounds challenging, but is at least clear: it could be worse, such as, "Imitative copying ... may involve second-order representation, in so far as the acts done by (and perhaps the intentions of) the model have to be translated from what is involved in doing them from the model's point of view (perceived from the imitator's point of view), into a metarepresentation permitting their performance from the imitator's point of view" (Whiten and Byrne 1991). What these, and many other definitions of imitation, share is that the imitator is assumed to need insight into the mind of the individual that is copied: that imitation relies on intentionality. But that is not really necessary. With the feeding skills of great apes in mind, I developed the "behavior parsing" model (Byrne 1999a, 2003b), a theory of imitation that is based instead on detecting *statistical regularities*, found within the variability of multiple performances of the same skilled sequence of action. No insight into the model's intentions is necessary for this to work.

Every execution of a motor act, however familiar and well-practiced it is, will differ slightly from others. Nevertheless, this variation is constrained—because if certain characteristics are missing or stray too far from their canonical form the act will fail to achieve its purpose. Watching a single performance will not betray these underlying constraints, but the statistical regularities of a repeated, goal-directed action can serve to reveal the organizational structure that lies behind it. Unweaned great apes spend most of each day within a few feet of their mothers, and (since their main nutrition still comes from milk) they have almost full-time leisure to watch any nearby activities, as well as learn about the structure of the local environment by their own exploration. By the time a young gorilla first begins to handle a plant like a nettle, for instance, at the rather late age of about 2 years because the stinging hairs discourage earlier attempts, it will have watched many hundreds of nettle plants being expertly processed by its mother.

Consider how a young gorilla might learn how to process stinging nettles from statistical regularities of observed behavior. Because great apes, like monkeys and people, have the ability to segment fluid action by recognizing component elements that are in their own repertoire, the mother's behavior will be perceived as a string of discrete elements, and each of these elements of action will be a familiar one that the young ape can already perform. At this time, the infant's repertoire of familiar elements of action derives: (i) from its innate manual capacities; (ii) from many hours of playing with environmental objects, such as plants and discarded debris of the mother's feeding; and (iii) from its own experience of feeding on other plants, perhaps ones simpler to process than nettles. Suppose that it also has some way of focusing on those particular sequences of its mother's action that are relevant to eating nettles: perhaps it has explored nettle plants and found that they are painful, yet puzzlingly the mother seems to enjoy interacting with them, making her nettle interactions intrinsically interesting. (Some such mechanism to focus learning on relevant action sequences will be essential for any "bottom-up" model of motor learning.) Then, because motor behavior is intrinsically variable, and plants also vary from individual to individual, the string of elements that the young gorilla sees when watching its mother eat nettles will differ each time. However, her starting point will always be a growing, intact nettle stem, and—because she is expert at this task—her final stage will always be the same, popping a neatly folded package of nettle leaves into her mouth. In between these points, variation will be particularly associated with non-critical parts of the performance, and certain aspects must necessarily be the same—or else, the result will simply be failure (see Figure 11.4).

With repeated watching, and a mind that tends automatically to extract the regularities from behavior that varies over time, a pattern will gradually begin to become apparent. The mother always makes a sweeping movement of one hand, held around a nettle stem, which is sometimes held in the other hand even though the plant is still attached to the ground, and this leaves a leafless stem protruding from the ground (which she does not eat). She always makes a twisting or rocking movement of the hands against each other, and immediately drops a number of leaf-petioles (which she does not eat) onto the ground. She always uses one hand to fold a bundle of leaf-blades protruding from the other hand, and holds down this folded bundle with her thumb. Moreover, these stages always occur in exactly the same order each time.

Statistical regularities, in behavior that is repeatedly observed, thereby mark out the minimal set of essential actions from the many others that occur during nettle eating but which are not crucial to success; and they reveal the correct order in which they must be arranged. Human babies, as young as 8 months old, can detect statistical regularities in repeatedly spoken strings of nonsense words (Saffran et al. 1996). Just such sensitivity to repeated orderings must therefore be active early in human development. The usefulness to the young gorilla of detecting regularities applies not only to the linear sequence of movements of each hand, but also to the way the hands and finger groups operate together. Where it is crucial for success that the hands are closely coordinated in space and time while doing different jobs, or where a single hand must be used for two purposes at once, then these conjunctions will recur in every string; conjunctions that are merely coincidental will not.

Other statistical regularities derive from the modular organization and hierarchical organization of skilled action. Whenever the operation of removing debris is performed (by opening the hand that holds nettle leaf-blades, and delicately picking out debris with the other hand), it occurs at particular places in the sequence. Also, on some occasions but not others, a section of the program sequence may be repeated twice or several times. For instance, the process described in words by "pull a nettle plant into range, strip leaves from its stem in a bimanually coordinated movement, then detach and drop the leaf-petioles" may be repeated several times, before the mother continues with removing

PARSING HIERARCHICAL STRUCTURES | 151

three different nettle plants

removing debris

detaching petioles

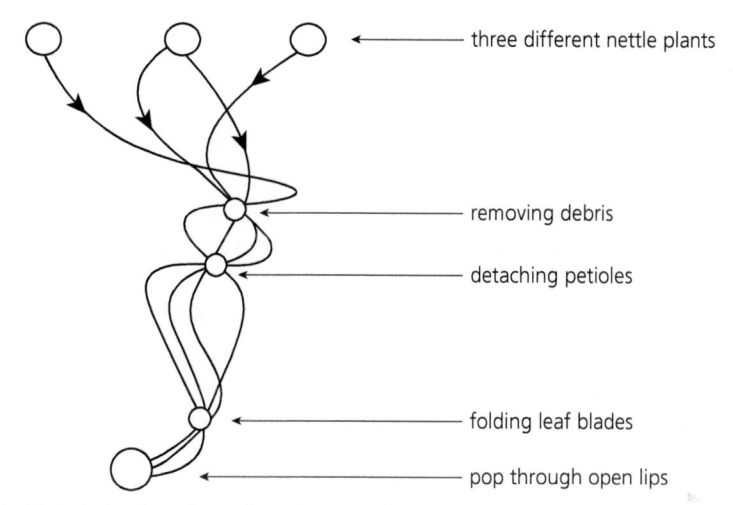

folding leaf blades

pop through open lips

Fig. 11.4 A visual analogy of nettle processing.

Imagine a 2D space in which each point represents the physical state of a nettle plant, and movement around the space represents a change of the plant's state: that's what I've tried to suggest with this analogy. Each nettle plant is slightly different to the others (some have flowers, some are unusually short, or tangled in other plants): thus the three different plants indicated are in somewhat different locations, at the top of the diagram. The operations applied to each plant by a skillful adult gorilla are represented by wavy lines down the page. The lines are naturally different to each other when the starting conditions differ; but also, even "repeated" actions are never exactly the same, so even the same plant would be processed slightly differently on different occasions or by different processors, which I've tried to represent by the two different lines from one of the starting points. Despite all this variation, however, the lines tend to cross each other at several, roughly similar places; that is because, unless the actions applied produced roughly the same results, the plant would never be suitable for eating. These states are represented by blobs, labeled with approximately the operations they correspond to. For instance, only when the leaf-blades have been detached from their petioles and wrapped in a single leaf, by means of folding, are they ready to be popped into the mouth without touching the lips: the last stage, at the bottom of the diagram.

Because of these constraints, some states of nettle plant recur time and time again: they have to for the job ever to be done. With a perceptual system specialized at extracting recurring, similar states from viewing repeated skilled behavior, those points are extracted "automatically"—and their nature and ordering are precisely what is needed to copy the process. Note that this analogy could just as well have been phrased in terms of extracting states as extracting actions: the action of detaching petioles comes before that of folding a leaf bundle, or the state of detached petioles comes before that of folded leaf bundle. Whether it is easier for observers to extract recurring actions or recurring transitions between states is an empirical question: either will serve to allow program-level imitation.

debris and folding the leaf-blades before eating. Subsections of the string of actions that are marked out in this way may be single elements, or as in this example a string of several elements. Both omission and repetition signal that some parts of the string are more tightly bound together than others, i.e., that they function as modules. Optional stages, like cleaning debris, occur between but not within modules; repetition of a sub-string gives evidence of a module used hierarchically as a subroutine, for example, iteration to accumulate a larger handful.

Further clues to modular structure are likely to be given by the distribution of pauses (occurring between but not within modules), and the possibility of smooth recovery from interruptions that occur between modules. Gorillas often pause for several seconds during the processing of a handful of plant material, in order to monitor the movements and actions of other individuals: they will naturally interrupt their feeding between modules rather than within them. Finally, a different module entirely may be substituted for part of the usual sequence. For instance, if one hand is required for postural support, then a normally bimanual process may need to be performed unimanually. If the substituted module is recognized as an already-familiar sequence, its substitution again reveals the latent structure. Eventually, it may be that a taxonomy of substitutable methods is built up.

All these statistical regularities are precisely what enabled us, the researchers, to discover the hierarchical nature of nettle processing by adult gorillas (Byrne and Byrne 1993; see Figure 11.5). In developing the *behavior parsing model*, I propose that the same information can be extracted and used by the apes themselves, and that this ability is what enables a young ape to perceive and copy the sequential, bimanually coordinated, hierarchical organization of complex skills from repeated watching of another. From what we know of living apes in the wild, it is entirely possible that non-human apes' capacity to parse behavior is limited to the visible domain of manual and bodily actions and thus not available in the auditory domain; in contrast, modern humans are routinely able to parse vocal material. (The bonobo Kanzi's apparent ability to parse human speech, when he responds correctly to words whose referent depends on the syntactical organization of a relative clause within a sentence, may cause this qualification to be relaxed at least for extensively human-reared apes: Savage-Rumbaugh et al. 1993. On current evidence, apes living under natural conditions have no such ability.) The great ape forte is predominantly in the manual domain, as convincingly demonstrated in the hundreds of ASL signs acquired by participants in "ape language" experiments (Gardner and Gardner 1969; Gardner et al. 1989).

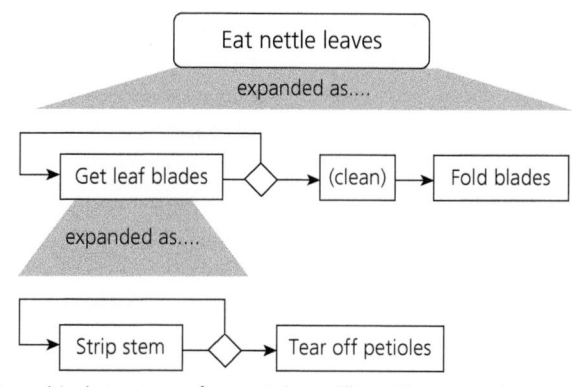

Fig. 11.5 Hierarchical structure of mountain gorilla nettle processing.

The evidence that nettle processing consists of two consecutive modules, rather than a single learnt sequence, is that (1) the first module ("Get leaf-blades") can be repeated several times, or not if it is not needed, and that (2) before the second module ("Fold blades") an optional cleaning stage may be inserted, or not if it is not needed. The evidence that the first of these modules is itself composed of two distinct modules is similar: the first ("Strip stem") may be iterated several times or not, before the second ("Tear off petioles") is begun. This structure is the *minimum* necessary to account for the observational evidence: in reality, the hierarchy may be considerably deeper and more elaborate.

Behavior parsing therefore enables the underlying hierarchical organization of planned behavior to be picked out—under certain circumstances. From the way the model works, the critical condition is that "multiple independent looks" are necessary. A single view of some skilled behavior that is unfamiliar in its organization will not result in a useful parsing, so seeing *multiple* samples of efficient behavior is required. The samples must be *independent*, so that there is information about the variance within the strings of perceived elements. That is because only by having sensitivity to the relative variability of elements can the process of behavior parsing locate the key (unvarying) elements. Thus, viewing a film clip of the same segment of skilled behavior would not serve to allow unfamiliar behavior to be parsed. Although we may well substantially overrate our everyday abilities (Bargh and Chartrand 1999), modern humans are not subject to this limitation, as was shown neatly in Baldwin's study, mentioned earlier, in which people viewing a single film clip of an activity reduced to fluorescent dot movements were able to parse the action into logical phases.

The behavior-parsing model therefore makes very clear predictions about when a great ape should and should not show program-level imitation, depending on how many independent sequences of planned action directed to the same goal it has been able to watch. Two or three clear presentations, as normally used in the experiments derived from child psychology and applied to apes, will simply not be enough, and indeed, great apes have typically failed such tests. Dozens, maybe hundreds of sequences are required, all from a skilled practitioner. This does not imply intense peering, and the natural upbringing of a wild great ape will automatically provide just the kind of experience that is necessary to learn skilled food processing, including tool making and tool use, from relatively casual observation of the mother's behavior. In order to use the hierarchical information about behavioral organization for program-level imitation, there must also be a reliable way of storing that information and converting it to action: a working memory. The fact that the hierarchical structures found in gorilla food processing are relatively "shallow" perhaps hints that gorilla working memory capacity is limited in size compared to that of humans.

Beyond imitation

This chapter has had to wade into the definitional technicalities of imitation, but I hope it has now become clear that there was a good reason for that: the aim of getting to the evolutionary origin of great ape insight. Imitation has long been considered to depend on representational understanding, and the coincidence that just those primate species showing insight in other domains also give evidence of imitation supports that linkage. But when we examine the way in which they learn by imitation—by behavior parsing—the implications become more powerful. The ability to parse the behavior of others, and thereby to learn some of their skills by observation—along with all the other individual and social ways of skill learning that are shared by many animals—is important for great apes. More than that, it is a fundamental attribute of human behavior, with implications far beyond the advantage of picking up tricky ways of food processing in an efficient way, however important that might be for a gorilla on a cold, wet Rwandan mountain. In Chapter 12 I will develop a view of how behavior parsing may have led to those forms of insightful understanding shown by our close relatives, as reviewed in earlier chapters, and which ultimately formed the basis for unique achievements of human cognition.

Chapter 12

A road map to insight

How it is we can think about why things happen

It is a truism to state that humans possess unique cognitive characteristics: imagining the future and making it happen, understanding and using modern technology, and above all using language to regulate and make sense of our social world. The aim of this final chapter is to chart a plausible pathway from the set of cognitive abilities that forms our shared heritage with monkeys and prosimian primates (and many other species of animal as well), to the unique abilities that characterize our species. The case I have built up over previous chapters is that the key to these modern human achievements is *insight*, that is, the ability to represent (or in everyday terms, to understand) how things and people work, what makes them tick—whether wristwatch or politician—and to compute new information from these mental representations, making decisions on the basis of predictions about the future. The argument in this chapter will be that, despite the fact that glimmerings of insight are evident in the social capacities of primates, the evolutionary origin of insight in the primate lineage was in dealing with the physical world. Specifically, the seminal development was behavior parsing: most likely an adaptation originally selected by increasing pressure on ancestors of the living great apes to feed more efficiently in the face of changing climate, but ultimately affecting all aspects of how an individual deals with the world.

Cognition without much insight

My starting point is the common cognitive heritage that we as humans share with primates in general: with monkeys, and in some cases also with prosimian primates. This can be seen as the springboard from which the cognitive capacities allowing insight evolved in one subsequent lineage. Whether primates in general are cognitively advanced over most other mammals is unimportant for my argument, though that claim has been defended by Michael Tomasello and Josep Call (1997). It could be argued that dolphins, parrots, crows, and elephants have abilities closer to those of humans than are the abilities of monkeys. (The evidence is less good for prosimians at present, so I will often refer specifically to monkeys, but this is not meant to suggest any great conviction of prosimian inabilities.)

But the point is, only primate evidence can tell us about the sequence of evolutionary changes in our direct line of descent. Using data from modern species, we can reconstruct the cognitive abilities of the primate population from which the great ape lineage, including humans, evolved—and doing so shows that this was already a species with remarkable cognitive capacities. As a rough and ready rule we can attribute to this ancestral species all those capacities that have been found to be shared widely among modern primate groups. What are they?

All primates come equipped with powerful mechanisms for learning, especially social learning. They follow other individuals' gaze, and can use this information to search more efficiently. They have the option of social learning, from being with others. In particular, places, objects, and actions are "primed" by seeing them in the context of a conspecific's behavior, making learning by exploration faster. Social learning of this kind allows rapid exploitation of others' knowledge and sometimes results in local traditions, though that brings the risk of conservatism. Monkeys and apes are able to exploit other individuals by forming temporary alliances or building long-term friendships of mutual help and support; sometimes they exploit others by deception, although that brings the risk of reprisal or counter-deception. In order to do all those things, it is clear that individuals of many species of primate are able to build up extensive social memories. Doing so not only involves recognizing a number of other individuals as such, accurately interpreting their demeanor, and remembering their idiosyncratic characteristics, but also responding appropriately to their kinship or affiliation patterns, their dominance ranks—relative to others as well as to the individual itself—and whether favors are owed to or owing from them. Experimental studies have shown that a monkey's social database is augmented even by overhearing the sounds of an out-of-sight interaction between third parties. To notice and remember such things, these primates must possess sophisticated perceptual systems, capable of distinguishing subtleties of behavior as well as an array of visual and auditory cues that allow them to distinguish dozens of individuals from each other and recognize them at a glance.

In many cases, the data on which that summary is based come from primate species that live in large social groups, like baboons, macaques, and chimpanzees. Moreover, primates that live in long-lasting social groups face an extra cognitive challenge compared to more solitary species. Conspecifics naturally present the most intense competition that an individual can face, yet if group living is critical for survival (principally to reduce predation risk, in the case of primates), any competition that disrupts it would have dire consequences. The theory that, in consequence, group living selected for greater intelligence in managing social relationships has been supported by evidence from mathematical analyses of brain sizes. In primates, and in several other kinds of mammal, those species that live in typically large groups also have a large neocortex in proportion to the rest of the brain. (Often, their overall brain size is larger as well, suggesting that neocortical enlargement cannot safely parasitize other brain systems with impunity.) Moreover, species with a larger neocortex have been found to commit more acts of deception, they more often show that they can and do learn socially, they tend to use tools more frequently, and they show more innovation. Of these abilities, the most impressive have been shown by the larger-bodied monkeys and apes; that makes sense, when the overall relationship between brain size and body size is remembered. The brain is energetically costly and remorseless in its energy demands, so that a brain that is large relative to the size of the body presents considerable survival risks. Already, this paints a picture of an animal with pretty impressive cognitive abilities. Yet not one of these things requires insight, in the sense of representational understanding and computing new information. Of course, individuals with insight—like ourselves—might well apply that insight to any of these tasks, and might as a result do them better. But each of the abilities I have sketched can be plausibly explained in simpler ways, and evidence for representational understanding of them by non-human primates is in most cases lacking.

Some primates have also been found to show considerable sophistication in interpreting the physical world: again, the evidence comes chiefly from larger-bodied monkeys and apes, though perhaps only because they are more intensively studied. Individuals of these species have been found to compute the geometry of others' viewpoints and plan their actions accordingly. Most impressively, their decision-making is not restricted to the immediate present. They can decide between alternatives that are not present and between out-of-sight locations which would take significant time to reach. In doing so, they sometimes take account of potential future competition; they can even model the effect of fruit ripening as a function of ambient temperature and sunshine. Thus, specifically in the spatial domain, non-human primates—and many other animals—do show a kind of insight, computing novel information from stored representations of reality. This "foraging insight" seems to be primitive, in that it is found in a much wider range of animals than just the primates, and is certainly not restricted to the great apes. Moreover, it does not seem associated with any more general insight, as we typically think of it—into how things work, why individuals do things, and all the other situations that we as humans seek to understand.

A forte of apes

The clue to the evolutionary origin of the capacity to apply representational understanding more generally, particularly in understanding other individuals (which we might call "social insight"), comes from data on great apes. That is not because apes are more social animals in any obvious way. Modern apes do not systematically differ from monkeys in social system, and in particular their social organizations do not seem to present greater cognitive challenge. Group size has been used as the metric of social challenge for species known to live in semi-permanent groups and whose reactions to others depend on knowing their identity. Many monkey species actually live in larger groups than apes. (If apes today have greater insight into their social companions, that itself may make their cognitive challenge greater—but it would be circular logic to use that as the explanation of how this came about.) Where apes do differ systematically is in their ability to develop impressively skilled feeding techniques, as pointed out in Chapter 9; and the capacity to do so is based on behavior parsing, as explained in Chapter 11.

Whereas the enhanced learning and social-networking abilities of all primates have generally been linked to the expansion of the neocortex, the brain area that is distinctively enlarged only in great apes—and especially in humans—is the cerebellum (Barton 2012; MacLeod et al. 2003). The cerebellum contains far more neurons than the neocortex; and for primates in general the two structures scale together, so all the correlates of the neocortical expansion—challenges of group living, frequency of use of deception, innovation, etc.—could equally be attributed to the cerebellum. The cerebellum's role, however, has traditionally been relegated to movement control and timing. But when he examined its neural tissue from a computational standpoint, David Marr came to a very different conclusion (Marr 1969). He used the superb diagrams of nineteenth-century anatomist Santiago Ramón y Cajal to show the network connectivity of the cerebellum, and modern electrophysiology to indicate how its synaptic connections functioned, and thereby tried to deduce the cerebellum's function from its engineering properties. Marr's conclusion was that the cerebellum was ideally designed for complex scheduling of elaborate actions: an ideal structure to allow development of the complex and hierarchically structured motor programs we saw in great ape feeding skills. Recently, Rob Barton has developed a strong case that it is rapid cerebellar enlargement in recent ape evolution that allows the ape-specific abilities in execution and understanding of behavioral sequences that I have identified as behavior parsing (Barton and Venditti 2014).

Behavior parsing gives the ability to "see below the surface" of behavior and detect the logical organization that produced it, by extracting statistical regularities in behavior that is seen many times. So far, I have treated this merely as a way of learning new skills by imitation, by parsing out and thereby copying the hierarchical organization of another's behavior, thus benefiting from the discoveries of companions and even individuals in past generations. But perhaps it was no coincidence that it was in the great apes—species already able to parse hierarchically organized behavior—that language and speech and all the other hallmarks of modern human cognition evolved. That is the contention of this final chapter: that behavior parsing has implications for other cognitive activities than just imitation, and is part of a fundamental process of interpreting and understanding complex behavior.

Parsing to "see" causes and intentions

In the behavior-parsing model, the process of parsing starts from observing skilled behavior, and requires no prior understanding of the physical cause-and-effect of the actions upon objects in the world, nor of the intentions of the demonstrator. Behavior parsing might, however, be a necessary step on the road to seeing the world in an intentional–causal way: a key to the origin of insight into mechanical causes and intentional actions.

Consider causation. When the behavior that is parsed involves operating with things in the environment, the physical changes that are seen to take place to those things will be parsed along with the actions applied to them. So, changes in the physical world will become linked to the sequence of action—statistically. If a certain sequence of actions regularly makes a certain result happen in the world, then in a statistical sense the actions caused the change. In this way, behavior parsing makes mechanical causes evident: as correlations. Of course, there is more to cause than correlation, but maybe that doesn't matter much for everyday purposes, or for evolution. A reliable correlation of this kind might be described as a "Pretty Good Cause," and only physicists or philosophers dealing with the fundamentals of matter may need to go much beyond it.

The fact is that, even for people, most things are seen as likely to happen because they or things very like them have happened often before under the same circumstances. The sun will rise tomorrow morning because it has been doing so for a long time at rather regular and predictable intervals; not flawless logic, but good enough. Any parent who has tried to answer a series of "Why?" questions from a young child will know how soon one gets out of one's depth with causation. You know the sort of thing: "Ok, mum, so you say that day and night are caused by the Earth going round the Sun, but why does it do that?" In fact, probing deeper into the physics of most everyday situations helps little with everyday living, and often does not provide a satisfying advance on cause-as-correlation. Replacing the pretty good correlational principle that things "always fall down, not up" with the concept of an omnipresent and invisible force field, gravity, let alone with an exchange of (totally undetectable) virtual graviton particles, leaves most people cold. In contrast, behavior parsing picks out the correlational structure of a changing environment quite well. *Cause-as-correlation* is a valuable everyday way of representing reality.

Consider intentionality. Any organization of behavior that is extracted by the parsing process will inevitably be set in a real-world context in which individuals are satisfied when they achieve valuable ends. This follows just because the individuals observed to be engaged in skillful action will only be doing so for biologically good and sufficient reasons. Typically, those individuals will be close associates or relatives of the observers, confronting much the same problems as they do themselves. So, associating a particular organizational structure with the typical, satisfying result of its performance will in many cases be an easy task. Indeed, the point of achieving that particular result is something the observer probably already understands. In this way, intended purpose is indicated statistically: by the usual result of successful performance. ("Unsuccessful" can also be identified statistically on the basis of visible behavior. It corresponds to those occasions when the individual is evidently not satisfied and re-does the action, in the same or another way, rather than moving on to another activity.) This means that, in principle, behavior parsing makes it possible to compute the prior intention of the other individual, the purpose of their behavior: by recognizing a behavior pattern that would achieve a goal that makes sense for them, if they themselves performed that pattern. An animal capable of behavior parsing, and who has some prior experience of behavior they observe in others and the environment they both inhabit, should therefore also be able to detect at least some of the intentions of those others—from their behavior alone.

As in the case of causation, the intentions extracted by behavior parsing are intentions in a weak sense of the term. Rather than an imagined mental state, intentions of this sort need be no more than the proper results of the normal behavior sequence. But just as with causation, such a "Pretty Good Intention" may be good enough for most everyday purposes: animals sensitive to *intention-as-result* won't be able to conceive of false belief and deliberate trickery, but they will be able to pick out the intended purposes of many everyday actions and social signals, and gain insight into the self and others sufficient to explain mirror self-recognition and empathy. Pretty good intentions still imply mentalizing, though falling short of theory-of-mind capacity as normally envisaged. Combined with the delicate and sophisticated manual control of action that we find in all the living apes, even this limited kind of intentional understanding should be sufficient for communication by means of gesture. And as we saw in Chapter 4, all the living great apes use elaborate systems of gestures for intentional communication.

Going beyond "pretty good" causes and intentions: Two kinds of insight

Behavior parsing is therefore not only a powerful way of parsing out the organization of others' (oft-repeated) actions, allowing efficient learning of causally structured plans, but it also gives "pretty good" versions of intention (as satisfying outcomes) and causality (as correlations). Nevertheless, it is clear that modern humans can go beyond this statistical kind of insight, to represent causality and intention in deeper ways, and compute with intentions as mental states and underlying causes of an abstract nature: as we'd say in everyday life, to have some real understanding of what's going on. The natural assumption might be to assume that the modern human version of insight "replaced" the merely statistical insights of behavior parsing in non-human great apes: I suggest, however, that that may not have been quite what happened.

I propose that insight evolved twice, and that both kinds of insight are available to adult modern humans (Figure 12.1). Non-human great apes, with their understanding of causation limited to correlations and their understanding of intentions limited to expected results, can do very well at non-verbal experimental tests of insight, because the tests normally rely on getting results, rather than showing understanding. To show understanding, the developmental psychologist Jean Piaget suggested that it is crucial to produce a correct verbal account of the situation, not just to get the correct answer. Thus, today, in developmental psychology there is a gulf between the ages at which children can show theory of mind by describing mental states, and those at which the same children show theory of mind by their observed behavior alone. I suggest that this apparent conflict will not be "resolved," because it accurately reflects two kinds of insight, gained by the child at different ages. First, the child becomes able to pass non-verbal tests of theory of mind, by using information derived statistically by behavior parsing: pretty good intentions. (And I predict the same will be found to apply to tests of cause-and-effect understanding.) Only much later, the child becomes able to reflect upon what she sees: a process described by Annette Karmiloff-Smith as "re-representation" (1993). Then, the child can at last explain what is going on in terms of invisible entities such as mental states.

This proposal makes the strong predictions, that:

- all non-verbal tests of theory of mind will eventually be "passed" by a non-human great ape; and
- all non-verbal tests of cause-and-effect logic will eventually be "passed" by a non-human great ape.

[Diagram: Proposed primate evolutionary path

- rapid learning in social contexts — 'Machiavellian' simian state
- 12Mya → ability to parse **organization** and **effects** from behavior; causal **attribution** and **planning** — from a need to acquire novel manual skills, in order to feed more efficiently; primary adaptation, in great ape ancestor
- 2Mya? 200Kya? → mentalizing — secondary, only in humans, based on language]

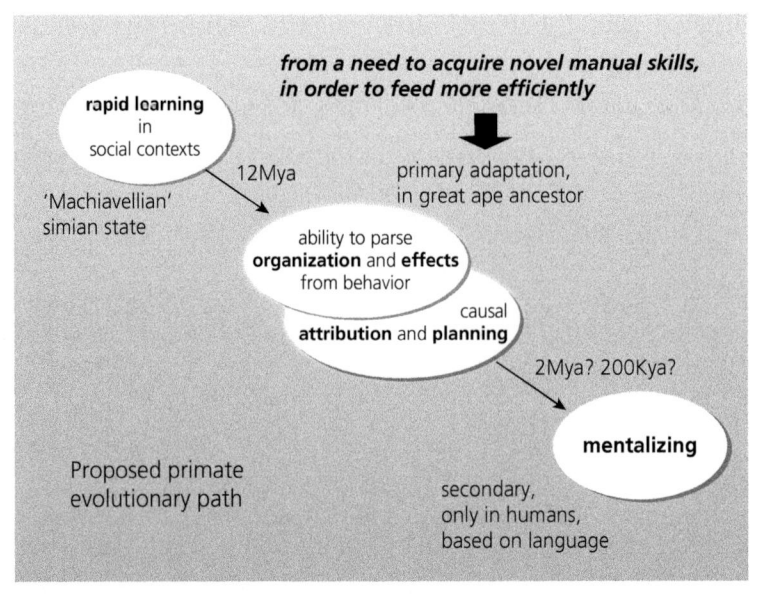

Fig. 12.1 Proposal for two kinds of insight into mental states.

On this hypothesis, the ability to compute and react to mental states of other individuals evolved twice: once in the ancestor we share with all living great apes, and once in the uniquely human line. The capacities that resulted were different, though both are useful. The starting point was a simian state, shared today with monkeys and other primates, in which particularly rapid learning was possible in social contexts—allowing smart-looking behavior of several kinds, such as cooperation and deception, but without insight into how these tactics worked. Rapid learning is a product of large neocortex size, and since the ancient apes were quite large primates, they would be particularly versatile in this "insight-less" intelligence. The first species on the human line to be able to show some real insight was the ancestor species that gave rise to all modern great apes (technically the "crown group ancestor"), at around 12 million years ago or somewhat before. Driven by a need to forage and process food more efficiently, in response to the drying climate that made conditions very difficult for the early apes, the ability to parse behavior evolved in this ancestor species, along with a much larger cerebellum. Parsing allowed the organization and effects of others' behavior to be discerned, and thereby copied, allowing much more efficient acquisition and transmission of skills, once they had been discovered by one individual. This species would therefore have shown many traditions of impressively complex behavior; in addition, individuals would have been able to make generally accurate attributions of how things worked, both in the physical realm and the psychological one, and shown relatively advanced planned behavior. Such attributions, however, would not have gone as far as false belief and its consequences, in social interactions, nor postulation of invisible causes in the physical realm, such as gravity. The extra level of complexity for these representations would require language, in the sense that we see it in modern humans, and that's why there are question marks on the diagram! Whether language in the full, modern sense was possible for the first *Homo* species around 2 million years ago, or in the first *Homo sapiens*, around 200 thousand years ago, or only at some other date such as that of European cave art, is simply not known. But the kind of mentalizing that adult humans are capable of today would not have been possible until after that magical date.

The second kind of insight, of course, must be parasitic on language, with its immense power and efficiency for representing and manipulating complex informational structures. And language most certainly developed only in the uniquely human line of evolution, the lineage leading to ourselves from the last common ancestor. It is frustrating that we cannot be much more precise than "in the uniquely human lineage": after all, the comparative evidence sketched in earlier chapters allows us to say that behavior parsing, and thus potentially a statistical sort of insight, evolved about 12 million years ago, the most recent date for the common ancestor of all the living great apes. But at least three different dates have been proposed for the evolution of language: about 2 million years ago, the date of the sudden increase in brain size associated with *Homo erectus*; about 200 thousand years ago, the date associated with the origin of modern *Homo sapiens* in southern Africa, as indicated by molecular chronology, and the first signs of practices today associated with culture, such as ochre mining and abstract notations; and about 40 thousand years ago, the date of the earliest cave paintings of southern Europe, which convince most researchers that their authors must have possessed modern human ways of representing abstract entities. Unfortunately, the choice among these hypotheses does not seem to be decidable on any current evidence, and different authors see quite different scenarios as the most plausible.

That uncertainty makes it difficult to identify what selective pressure resulted in the evolution of language in its modern form. According to the three date proposals, language would have originated in very different sorts of animal, living under very different conditions. Moreover, even this way of talking of "an origin" of language is almost certainly simplistic. Far more likely, it proceeded in steps: the first "proto-language" might have been based on gesture, a gestural language with the reference and syntax that are missing from the gestural communication of living apes; or on speech without syntax, a flexible vocabulary of vocal signals that could be developed as needed, by hominins with the ability to acquire new vocalizations that is missing from the vocal communication of living apes. (To me, the former seems more likely: "ape language" studies show that great apes can readily use gestures referentially, and studies of skilled food processing show they readily develop structures of action that match the hierarchies of phrase-structure grammar.) On either speculation, more than one "origin" date is required. There is abundant evidence that speech is a biological property of humans (Lenneberg 1968), but whether that is also true of language is less clear (though strongly argued by some: Pinker 1994). Alternatively, it may still be that the set of conventions that makes up the universal grammar of language was an invention, just as undoubtedly were the words of each of the thousands of languages known today.

Never mind! All that is needed to underpin my theory of two kinds of insight is that language developed very much more recently than the behavior-parsing abilities common to all the great apes, and there is no doubt about that.

Making sense of apes

In several ways, this theory meshes well with modern views. It is founded, for instance, on developments in the motor domain. Biological anthropologists have long pointed to the importance of prehensile hands in the primate lineage, and especially the ability of apes to support precision grips (Christel 1993; Napier 1961), relating the latter development to the more extensive representation of the hands in ape motor cortex (Deacon 1997a). From this starting point, the idea of behavior parsing explains the superior abilities of apes—with their disproportionately enlarged cerebellum—to learn complex, novel skills by program-level imitation. Behavior parsing forms a crucial bridge to a simple, statistical form of insight into the mechanical and social domains. In "statistical insight," causality is detected as results that regularly recur in the same circumstances, and intentions are detected as outcomes that appear to satisfy the observed actor. Thus, according to the theory developed here, an ability that evolved in the technical realm as a way of feeding more efficiently was secondarily applied to social entities. At first, it was perhaps applied only to social entities that operated on the physical world, as in the case of a mother ape processing plant food or making tools for insect fishing; later, the same cognitive subsystems were applied more generally to detect (pretty good) social causes and personal goals in purely social situations, and to detect (pretty good) mechanical causes for everyday happenings in the purely physical world. The result was—in great apes generally—an incomplete but still useful kind of insight into others' intentions and physical causation.

Great apes construct novel programs of action which follow a hierarchical pattern of embedding of subroutines, closely reminiscent of phrase-structure grammar in language. But when it comes to communication, they show no trace of syntactical structure. Thus, apes may build novel, hierarchically organized actions, but they seem to have no concept of the power of hierarchical organization more generally. Even when artificially scaffolded with many of the building blocks of modern human cognition, as learnt in "ape language" projects, no evidence is seen that the subjects realize the potential that the system of communicating they've been taught would have if put to productive use: the *idea* of communicating intentions. Great apes use their gestures intentionally to modify and influence the behavior of others, but they remain restricted (in nature) to the gestures conferred on them by biology. They show no sign that they understand how useful it might be to propose and introduce new labels, for example to refer to objects and individuals: the *idea* of reference is missing. Great apes sometimes acquire behaviors that function to deceive others, as do all groups of primate, and they go further in occasionally computing new ways of profiting by deliberately ensuring the ignorance of others. Even so, they show no sign of understanding the *idea* that they have sometimes been deceived by others, or harboring motives of revenge for deceit. Great apes can understand their reflection in a mirror, showing that they have developed some concept of themselves as an entity: but they don't go on to show embarrassment or pride in appearance, indicating the *idea* of how they look to others. In circumstances where learning individually is very difficult, such as learning to handle stone tools, apes have a range of behavioral traits that aid another's learning, a process which might be called functional teaching (in this case, the best evidence comes from chimpanzees): allowing an infant to take over a half-processed food, or to take over a prized stone tool. Very rarely, their behavior appears to match true pedagogy, such as modeling the necessary action with exaggerated slowness. Yet if chimpanzees or other apes understood the *idea* of pedagogy and the value of ability-specific teaching, such targeted helping to learn would surely occur frequently enough to be observed regularly, which it does not.

We may never know in which circumstances the adaptive benefits of some kind of language drove its evolution. The benefits come both directly, in enhanced communication, and indirectly in allowing robust mental representations that can be more easily manipulated and held. Cooperation in larger social groups has often been proposed; that might include the benefits of retrospectively reinterpreting malicious deceit as accident or misunderstanding. Alternatively, the pressure to develop better and better tools may have favored intentional pedagogy, driving a technical form of language as the vanguard of everyday speech. But whatever it was, it could only have happened in a species with the mental capacities of the living great apes.

Tailpiece: A heretical thought

This chapter has argued that an understanding of planned behavior, in terms of hierarchically organized structure that can be copied, with causality approximated by correlation and purpose approximated by normal results, can result from a mechanistic process of behavioral analysis. Such abilities are still seen in young children, as they are in other species of great ape. There is no need for any mentalizing about the mind of the observed party, or mental representation of the real causal structure of physical system.

Adult humans, of course, can and do represent causes and intentions. In particular, we explain (away) our actions, on grounds of our beliefs, false or otherwise; we teach our children by explaining that one thing causes another or that some people have different beliefs to ourselves, and so on. But do these retrospective, verbal accounts actually correspond to causal mental states that generate our behavior when we are not explaining anything? Humans are remarkably reluctant to accept that much of our everyday behavior might be an automatic product of mental processes of which we are unaware (Bargh and Chartrand 1999), but I think this should be seriously considered for the case of theory of mind. The great ape system of "statistical insight" may not be put away as a childish thing: it might be crucial to our everyday adult lives.

Most of us are able to function effortlessly in a complex social world, without ever seeing this as a hard task or having to think much about it. There are two possibilities for why that is. On the one hand, it may be that calculations about others' mental states have a causal role, and that the normal process of "automatization" with extensive practice has rendered the mental-state calculations fast and efficient, to the point when they can only be made conscious by careful deliberation. The obvious analogy is to the skill learning involved in driving a car: it's quite hard for an experienced driver to work out what precisely they do first with their eyes and feet when turning across traffic, but they can do it safely and effortlessly; in contrast, a learner driver may know the proper sequence of acts, but not be so good at executing them.

The heretical alternative is that unconscious processes—based on extraction of statistical regularities, allowing us to parse behavior, as can any other great ape—actually *cause* most of our everyday social behavior and our interactions with the world of objects. Mentalizing is not a replacement of statistical insight, it is an add-on. (The developmental psychologist Ian Apperly has come to a similar conclusion from a different standpoint: Apperly and Butterfill 2009.) On this view, mentalizing is used more sparingly and for different purposes: everyday stuff is handled by the phylogenetically ancient process.

Cases where we must and do use our mentalizing skills include teaching, when we explain processes or people to a child, and making excuses, when we retrospectively construe our deceitful behavior in a way very different to what we know to be accurate. Any such process of verbal (mis)construal is likely to be a function of language ability, and so must be recent in human evolution; but the everyday behavioral capacities that we attribute to "theory of mind" are perhaps shared with non-human great apes, though they cannot explain and discuss their actions as we can.

This heretical possibility of course depends on the theory, already argued, that the ability to deal with the mental states of others evolved twice, as suggested in Figure 12.1. The first process, based on the behavior-parsing abilities we share with other living great apes, is a statistical one, in which intentions-as-results are extracted from frequently observed behavior. Since humans have much larger cerebellum and neocortex than any of the non-human great apes, we are likely to be better at this than they are, and less limited by working-memory capacity. The second evolved more recently and uniquely in the human lineage; it requires language, and allows us to represent intentions formally and use them to compute counterfactual "what if" scenarios, including deceit by deliberate misrepresentation and teaching by understanding a learner's knowledge deficiencies. Most of the time, we rely on process #1, which is fast, efficient, and unconscious; when pushed, we can fall back onto the slower, but logically defensible and verbally explicable process #2. Perhaps, then, much of our everyday social lives is regulated and driven by capacities shared by our ape relatives, even though we alone can express these things in words!

References

Allman, J M, Hakeem, A, Erwin, J M, Nimchinsky, E, and Hof, P (2001), "The anterior cingulate cortex. The evolution of an interface between emotion and cognition," *Annals of the New York Academy of Sciences*, **935**, 107–17.

Allman, J M, Tetreault, N A, Hakeem, A Y, Manaye, K F, Semendeferi, K, Erwin, J M, Park, S, Goubert, V, and Hof, P R (2010), "The von Economo neurons in frontoinsular and anterior cingulate cortex in great apes and humans," *Brain Structure and Function*, **214** (5–6), 495–517.

Anderson, J R (1984), "Monkeys with mirrors: Some questions for primate psychology," *International Journal of Primatology*, **5**, 81–98.

Anderson, J R, Gillies, A, and Lock, L C (2010), "*Pan* thanatology," *Current Biology*, **20** (8), R349–51.

Anderson, J R, Myowa-Yamakoshi, M, and Matsuzawa, T (2004), "Contagious yawning in chimpanzees," *Proceedings of the Royal Society of London B: Biology Letters Supplement*, **271**, S468–70.

Apperly, I and Butterfill, S A (2009), "Do humans have two systems to track beliefs and belief-like states?," *Psychological Review*, **116**, 953–70.

Arnold, K and Barton, R A (2001), "Postconflict behavior of spectacled leaf monkeys (*Trachypithecus obscurus*). II. Contact with third parties," *International Journal of Primatology*, **22**, 267–86.

Balda, R P and Kamil, A C (1992), "Long-term spatial memory in Clark's nutcracker, *Nucifraga columbiana*," *Animal Behaviour*, **44** (4), 761–9.

Baldwin, D, Andersson, A, Saffran, J, and Meyer, M (2008), "Segmenting dynamic human action via statistical structure," *Cognition*, **106** (3), 1382–407.

Ban, S D, Boesch, C, and Janmaat, K R L (2014), "Tai chimpanzees anticipate revisiting high-valued fruit trees from further distances," *Animal Cognition*, **17** (6), 1353–64.

Bargh, J A and Chartrand, T L (1999), "The unbearable automaticity of being," *American Psychologist*, **54**, 462–79.

Barton, R A (1998), "Visual specialization and brain evolution in primates," *Proceedings of the Royal Society of London B*, **265**, 1933–7.

Barton, R A (2006), "Primate brain evolution: Integrating comparative, neurophysiological, and ethological data," *Evolutionary Anthropology*, **15**, 224–36.

Barton, R A (2012), "Embodied cognitive evolution and the cerebellum," *Philosophical Transactions of the Royal Society of London B: Biological Sciences*, **367** (1599), 2097–107.

Barton, R A and Dunbar, R I M (1997), "Evolution of the social brain," in A Whiten and R W Byrne (eds), *Machiavellian Intelligence II: Extensions and Evaluations* (Cambridge: Cambridge University Press), 240–63.

Barton, R A and Venditti, C (2014), "Rapid evolution of the cerebellum in humans and other great apes," *Current Biology*, **24**, 2440–4.

Bates, L A, Handford, R, Lee, P C, Njiraini, N, Poole, J H, Sayialel, K, Sayialel, S, Moss, C J, and Byrne, R W (2010), "Why do African elephants (*Loxodonta africana*) simulate oestrus? An analysis of longitudinal data," *PLoS ONE*, **5** (3).

Bates, L A, Lee, P C, Njiraini, N, Poole, J H, Sayialel, K, Sayialel, S, Moss, C J, and Byrne, R W (2008), "Do elephants show empathy?," *Journal of Consciousness Studies*, **15**, 204–25.

Bates, L A, Sayialel, K N, Njiraini, N, Moss, C J, Poole, J H, and Byrne, R W (2007), "Elephants classify human ethnic groups by odor and garment color," *Current Biology*, **17** (22), 1938–42.

Beck, B B (1980), *Animal Tool Behaviour* (New York: Garland Press).

Benhamou, S (1996), "No evidence for cognitive mapping in rats," *Animal Behaviour*, **52**, 201–12.

Boesch, C (1991), "Teaching among wild chimpanzees," *Animal Behaviour*, **41**, 530–2.

Boesch, C (1996), "The emergence of cultures among wild chimpanzees," in W G Runciman, J Maynard-Smith, and R I M Dunbar (eds), *Evolution of Social Behaviour Patterns in Monkeys and Man* (London: The British Academy), 251–68.

Boesch, C and Boesch, H (1984), "Mental map in wild chimpanzees: An analysis of hammer transports for nut cracking," *Primates*, **25**, 160–70.

Boesch, C and Boesch, H (1990), "Tool use and tool making in wild chimpanzees," *Folia Primatologica*, **54**, 86–99.

Boesch, C and Boesch-Achermann, H (2000), *The Chimpanzees of the Taï Forest: Behavioural Ecology and Evolution* (Oxford: Oxford University Press).

Bovet, D and Washburn, D A (2003), "Rhesus macaques (*Macaca mulatta*) categorize unknown conspecifics according to their dominance relations," *Journal of Comparative Psychology*, **117**, 400–5.

Boysen, S T, Berntson, G G, Mannan, M B, and Cacioppo, J T (1996), "Quantity based interference and symbolic representations in chimpanzees (*Pan troglodytes*)," *Journal of Experimental Psychology: Animal Behavior Processes*, **22** (1), 76–86.

Brewer, S and McGrew, W C (1990), "Chimpanzee use of a tool-set to get honey," *Folia Primatologica*, **54**, 100–4.

Brooks, R and Meltzoff, A N (2015), "Connecting the dots from infancy to childhood: A longitudinal study connecting gaze following, language, and explicit theory of mind," *Journal of Experimental Child Psychology*, **130**, 67–78.

Brothers, L (1990), "The social brain: A project for integrating primate behavior and neurophysiology in a new domain," *Concepts in Neuroscience*, **1**, 27–51.

Brüne, M, Ribbert, H, and Schiefenhövel, W (2003), *The Social Brain* (Chichester, West Sussex: John Wiley).

Bugnyar, T (2002), "Observational learning and the raiding of food caches in ravens, *Corvus corax*: Is it 'tactical' deception?," *Animal Behaviour*, **64**, 185–95.

Bugnyar, T (2007), "An integrative approach to the study of 'theory-of-mind'-like abilities in ravens," *Japanese Journal of Animal Psychology*, **57**, 15–27.

Bugnyar, T and Heinrich, B (2005), "Ravens, *Corvus corax*, differentiate between knowledgeable and ignorant competitors," *Proceedings of the Royal Society of London B: Biological Sciences*, **272** (1573), 1641–6.

Bugnyar, T and Heinrich, B (2006), "Pilfering ravens, *Corvus corax*, adjust their behaviour to social context and identity of competitors," *Animal Cognition*, **9** (4), 369–76.

Bugnyar, T, Stowe, M, and Heinrich, B (2004), "Ravens, *Corvus corax*, follow gaze direction of humans around obstacles," *Proceedings of the Royal Society of London B: Biological Sciences*, **271** (1546), 1331–6.

Busse, C D (1976), "Do chimpanzees hunt cooperatively?," *Nature*, **112**, 767–70.

Byrne, R W (1979), "Memory for urban geography," *Quarterly Journal of Experimental Psychology*, **31**, 147–54.

Byrne, R W (1981), "Distance calls of Guinea baboons (*Papio papio*) in Senegal: An analysis of function," *Behaviour*, **78**, 283–312.

Byrne, R W (1993), "A formal notation to aid analysis of complex behaviour: Understanding the tactical deception of primates," *Behaviour*, **127**, 231–46.

Byrne, R W (1994), "The evolution of intelligence," in P J B Slater and T R Halliday (eds), *Behaviour and Evolution* (Cambridge: Cambridge University Press), 223–65.

Byrne, R W (1995a), *The Thinking Ape: Evolutionary Origins of Intelligence* (Oxford: Oxford University Press).

Byrne, R W (1995b), "Primate cognition: Comparing problems and skills," *American Journal of Primatology*, **37**, 127–41.

Byrne, R W (1996), "The misunderstood ape: Cognitive skills of the gorilla," in A E Russon, K A Bard, and S T Parker (eds), *Reaching into Thought; the Minds of the Great Apes* (Cambridge: Cambridge University Press), 111–30.

Byrne, R W (1997a), "What's the use of anecdotes? Attempts to distinguish psychological mechanisms in primate tactical deception," in R W Mitchell, N S Thompson, and L Miles (eds), *Anthropomorphism, Anecdotes, and Animals: The Emperor's New Clothes?* (New York: SUNY Press, Biology and Philosophy), 134–50.

Byrne, R W (1997b), "The technical intelligence hypothesis: An additional evolutionary stimulus to intelligence?," in A Whiten and R W Byrne (eds), *Machiavellian Intelligence II: Extensions and Evaluations* (Cambridge: Cambridge University Press), 289–311.

Byrne, R W (1998), "Imitation: The contributions of priming and program-level copying," in S Braten (ed.), *Intersubjective Communication and Emotion in Early Ontogeny* (Studies in emotion and social interaction; Cambridge: Cambridge University Press), 228–44.

Byrne, R W (1999a), "Imitation without intentionality. Using string parsing to copy the organization of behaviour," *Animal Cognition*, **2**, 63–72.

Byrne, R W (1999b), "Cognition in great ape ecology. Skill-learning ability opens up foraging opportunities," *Symposia of the Zoological Society of London*, **72**, 333–50.

Byrne, R W (1999c), "Object manipulation and skill organization in the complex food preparation of mountain gorillas," in S T Parker, R W Mitchell, and H L Miles (eds), *The Mentality of Gorillas and Orangutans* (Cambridge: Cambridge University Press), 147–59.

Byrne, R W (2000a), "How monkeys find their way. Leadership, coordination, and cognitive maps of African baboons," in S Boinski and P Garber (eds), *On the Move: How and Why Animals Travel in Groups* (Chicago: University of Chicago Press), 491–518.

Byrne, R W (2000b), "Is consciousness a useful scientific term? Problems of 'animal consciousness'," *Vlaams Diergeneeskundig Tijdschrift*, **69**, 407–11.

Byrne, R W (2001), "Clever hands: The food processing skills of mountain gorillas," in M M Robbins, P Sicotte, and K J Stewart (eds), *Mountain Gorillas. Three Decades of Research at Karisoke* (Cambridge: Cambridge University Press), 293–313.

Byrne, R W (2002b), "Imitation of complex novel actions: What does the evidence from animals mean?," *Advances in the Study of Behavior*, **31**, 77–105.

Byrne, R W (2003a), "Novelty in deception," in K Laland and S Reader (eds), *Animal Innovation* (Oxford: Oxford University Press), 237–59.

Byrne, R W (2003b), "Imitation as behaviour parsing," *Philosophical Transactions of the Royal Society of London B*, **358**, 529–36.

Byrne, R W (2004), "The manual skills and cognition that lie behind hominid tool use," in A E Russon and D R Begun (eds), *Evolutionary Origins of Great Ape Intelligence* (Cambridge: Cambridge University Press), 31–44.

Byrne, R W (2005), "Detecting, understanding, and explaining animal imitation," in S Hurley and N Chater (eds), *Perspectives on Imitation: From Mirror Neurons to Memes* (Cambridge, MA: MIT Press), 255–82.

Byrne, R W (2007), "Culture in great apes: Using intricate complexity in feeding skills to trace the evolutionary origin of human technical prowess," *Philosophical Transactions of the Royal Society (B)*, **362**, 577–85.

Byrne, R W and Bates, L A (2006), "Why are animals cognitive?," *Current Biology*, **16**, R445–48.

Byrne, R W and Byrne, J M E (1991), "Hand preferences in the skilled gathering tasks of mountain gorillas (*Gorilla g. beringei*)," *Cortex*, **27**, 521–46.

Byrne, R W and Byrne, J M E (1993), "Complex leaf-gathering skills of mountain gorillas (*Gorilla g. beringei*): Variability and standardization," *American Journal of Primatology*, **31**, 241–61.

Byrne, R W and Corp, N (2004), "Neocortex size predicts deception rate in primates," *Proceedings of the Royal Society of London B: Biological Sciences*, **271**, 1693–9.

Byrne, R W, Corp, N, and Byrne, J M E (2001a), "Manual dexterity in the gorilla: Bimanual and digit role differentiation in a natural task," *Animal Cognition*, **4**, 347–61.

Byrne, R W, Corp, N, and Byrne, J M E (2001b), "Estimating the complexity of animal behaviour: How mountain gorillas eat thistles," *Behaviour*, **138**, 525–57.

Byrne, R W, Hobaiter, C, and Klailova, M (2011), "Local traditions in gorilla manual skill: Evidence for observational learning of behavioral organization," *Animal Cognition*, **14** (5), 683–93.

Byrne, R W and Rapaport, L G (2011), "What are we learning from teaching?," *Animal Behaviour*, **82** (5), 1207–11.

Byrne, R W and Russon, A E (1998), "Learning by imitation: A hierarchical approach," *Behavioral and Brain Sciences*, **21**, 667–721.

Byrne, R W, Sanz, C M, and Morgan, D B (2013), "Chimpanzees plan their tool use," in C M Sanz, C Boesch, and J Call (eds), *Tool Use in Animals: Cognition and Ecology* (Cambridge: Cambridge University Press), 48–63.

Byrne, R W and Stokes, E J (2002), "Effects of manual disability on feeding skills in gorillas and chimpanzees: A cognitive analysis," *International Journal of Primatology*, **23**, 539–54.

Byrne, R W and Tanner, J E (2006), "Gestural imitation by a gorilla: Evidence and nature of the phenomenon," *International Journal of Psychology & Psychological Therapy*, **6**, 215–31.

Byrne, R W and Whiten, A (1985), "Tactical deception of familiar individuals in baboons (*Papio ursinus*)," *Animal Behaviour*, **33**, 669–73.

Byrne, R W and Whiten, A (1988), *Machiavellian Intelligence: Social Expertise and the Evolution of Intellect in Monkeys, Apes and Humans* (Oxford: Clarendon Press).

Byrne, R W and Whiten, A (1990), "Tactical deception in primates: The 1990 database," *Primate Report*, 27, 1–101.

Byrne, R W and Whiten, A (1991), "Computation and mindreading in primate tactical deception," in A Whiten (ed.), *Natural Theories of Mind* (Oxford: Basil Blackwell), 127–41.

Byrne, R W and Whiten, A (1992), "Cognitive evolution in primates: Evidence from tactical deception," *Man*, 27, 609–27.

Byrne, R W and Whiten, A (1997), "Machiavellian intelligence," in A Whiten and R W Byrne (eds), *Machiavellian Intelligence II: Extensions and Evaluations* (Cambridge: Cambridge University Press), 1–23.

Caldwell, M C and Caldwell, D K (1966), "Epimeletic (care-giving) behavior in Cetacea," in K S Norris (ed.), *Whales, Dolphins and Porpoises* (Berkeley: University of California Press), 755–89.

Call, J (2001), "Body imitation in an enculturated orangutan (*Pongo pygmaeus*)," *Cybernetics & Systems*, 32, 97–119.

Call, J and Tomasello, M (2007), *The Gestural Communication of Apes and Monkeys* (Hillsdale, NJ: Lawrence Erlbaum Associates).

Call, J and Tomasello, M (2008), "Does the chimpanzee have a theory of mind? 30 years later," *Trends in Cognitive Sciences*, 12, 187–92.

Caro, T M (1980), "Predatory behaviour in domestic cat mothers," *Behaviour*, 74, 128–47.

Caro, T M (1994), *Cheetahs of the Serengeti Plains: Grouping in an Asocial Species* (Chicago: University of Chicago Press).

Caro, T M and Hauser, M D (1992), "Is there teaching in non-human animals?," *Quarterly Review of Biology*, 67, 151–74.

Cartmill, E A and Byrne, R W (2007), "Orangutans modify their gestural signaling according to their audience's comprehension," *Current Biology*, 17 (15), 1345–8.

Cartmill, E A and Byrne, R W (2010), "Semantics of primate gestures: Intentional meanings of orangutan gestures," *Animal Cognition*, 13, 793–804.

Chance, M R A and Mead, A P (1953), "Social behaviour and primate evolution," *Symposia of the Society of Experimental Biology*, 7, 395–439.

Cheney, D L and Seyfarth, R M (1985), "Social and nonsocial knowledge in vervet monkeys," *Philosophical Transactions of the Royal Society of London B*, 308, 187–201.

Cheney, D L and Seyfarth, R M (1986), "The recognition of social alliances by vervet monkeys," *Animal Behaviour*, 34 (6), 1722.

Cheney, D L and Seyfarth, R M (1990a), *How Monkeys See the World: Inside the Mind of Another Species* (Chicago: University of Chicago Press).

Cheney, D L and Seyfarth, R M (1990b), "Attending to behaviour versus attending to knowledge: Examining monkeys' attribution of mental states," *Animal Behaviour*, 40, 742–53.

Christel, M I (1993), "Grasping techniques and hand preferences in Hominoidea," in H Preuschoft and D J Chivers (eds), *Hands of Primates* (New York: Springer Verlag), 91–108.

Christel, M I and Fragaszy, D (2000), "Manual function in *Cebus apella*. Digital mobility, preshaping, and endurance in repetitive grasping," *International Journal of Primatology*, 21 (4), 697–719.

Clayton, N S and Dickinson, A (1998), "Episodic-like memory during cache recovery by scrub jays," *Nature*, **395**, 272–8.

Clutton-Brock, J (1995), "Origins of the dog: Domestication and early history," in J Serpell (ed.), *The Domestic Dog: Its Evolution, Behaviour and Interactions with People* (Cambridge: Cambridge University Press), 7–20.

Cochet, H and Byrne, R W (2014), "Complexity in animal behaviour: Towards common ground," *Acta Ethologica*, **18**, 337–41.

Collins, D A and McGrew, W C (1987), "Termite fauna related to differences in tool-use between groups of chimpanzees (*Pan troglodytes*)," *Primates*, **28** (4), 457–71.

Cords, M (1997), "Friendships, alliances, reciprocity and repair," in A Whiten and R W Byrne (eds), *Machiavellian Intelligence II: Extensions and Evaluations* (Cambridge: Cambridge University Press), 24–49.

Corp, N and Byrne, R W (2002a), "The ontogeny of manual skill in wild chimpanzees: Evidence from feeding on the fruit of *Saba florida*," *Behaviour*, **139**, 137–68.

Corp, N and Byrne, R W (2002b), "Leaf processing of wild chimpanzees: Physically defended leaves reveal complex manual skills," *Ethology*, **108**, 1–24.

Crawford, M P (1937), "The cooperative solving of problems by young chimpanzees," *Comparative Psychology Monographs*, **14** (Serial Number 68).

Crockford, C, Wittig, R M, Seyfarth, R M, and Cheney, D L (2007), "Baboons eavesdrop to deduce mating opportunities." Animal Behaviour **73**, 885–890.

Crockford, C, Wittig, R M, Mundry, R, and Zuberbühler, K (2012), "Wild chimpanzees inform ignorant group members of danger," *Current Biology*, **22** (2), 142–6.

Cunningham, E and Janson, C H (2007), "Integrating information about location and value of resources by white-faced saki monkeys (*Pithecia pithecia*)," *Animal Cognition*, **10** (3), 293–304.

Custance, D M, Whiten, A, and Bard, K A (1995), "Can young chimpanzees (*Pan troglodytes*) imitate arbitrary actions? Hayes & Hayes (1952) revisited," *Behaviour*, **132**, 11–12.

Dally, J M, Emery, N J, and Clayton, N S (2004), "Cache protection strategies by western scrub-jays (*Aphelocoma californica*): Hiding food in the shade," *Proceedings of the Royal Society of London B: Biological Sciences*, **271** Suppl 6, S387–90.

Dally, J M, Emery, N J, and Clayton, N S (2005), "Cache protection strategies by western scrub-jays, Aphelocoma californica: Implications for social cognition," *Animal Behaviour*, **70** (6), 1251–63.

Dally, J M, Emery, N J, and Clayton, N S (2006), "Food-caching western scrub-jays keep track of who was watching when," *Science*, **312**, 1662–5.

Dally, J M, Emery, N J, and Clayton, N S (2010), "Avian theory of mind and counter espionage by food-caching western scrub-jays (*Aphelocoma californica*)," *European Journal of Developmental Psychology*, **7** (1), 17–37.

Dasser, V (1988), "Mapping social concepts in monkeys," in R W Byrne and A Whiten (eds), *Machiavellian Intelligence: Social Expertise and the Evolution of Intellect in Monkeys, Apes and Humans* (Oxford: Clarendon Press), 85–93.

Dawkins, R and Krebs, J R (1978), "Animal signals: Information or manipulation?," in J R Krebs and N B Davies (eds), *Behavioural Ecology: An Evolutionary Approach* (Oxford: Blackwell Scientific Publications), 282–309.

Deacon, T W (1997a), "What makes the human brain different?," *Annual Review of Anthropology*, **26**, 337–57.

Deacon, T W (1997b), *The Symbolic Species: The Co-evolution of Language and the Brain* (New York: W W Norton and Company).

Deaner, R O, van Schaik, C P, and Johnson, V (2006), "Do some taxa have better domain-general cognition than others? A meta-analysis of nonhuman primate studies," *Evolutionary Psychology*, **4**, 149–96.

de Waal, F B M (1982), *Chimpanzee Politics* (London: Jonathan Cape).

de Waal, F B M (1986), "Deception in the natural communication of chimpanzees," in R W Mitchell and N S Thompson (eds), *Deception: Perspectives on Human and Non-human Deceit* (Albany: State University of New York State).

de Waal, F B M (1991), "Complementary methods and convergent evidence in the study of primate social cognition," *Behaviour*, **118**, 297–320.

de Waal, F B M (2008), "Putting the altruism back into altruism: The evolution of empathy," *Annual Review of Psychology*, **59**, 279–300.

de Waal, F B M and Aureli, F (1996), "Consolation, reconciliation, and a possible cognitive difference between macaque and chimpanzee," in A E Russon, K A Bard, and S T Parker (eds), *Reaching into Thought: The Minds of the Great Apes* (Cambridge: Cambridge University Press), 80–110.

de Waal, F B M and van Roosmalen, A (1979), "Reconciliation and consolation among chimpanzees," *Behavioral Ecology and Sociobiology*, **5**, 55–6.

Di Fiore, A and Suarez, S A (2007), "Route-based travel and shared routes in sympatric spider and woolly monkeys: Cognitive and evolutionary implications," *Animal Cognition*, **10** (3), 317–29.

Douglas-Hamilton, I, Bhalla, S, Wittemyer, G, and Vollrath, F (2006), "Behavioural reactions of elephants towards a dying and deceased matriarch," *Applied Animal Behaviour Science*, **100** (1–2), 87–102.

Dunbar, R I M (1988), *Primate Social Systems* (London: Croom Helm).

Dunbar, R I M (1991), "Functional significance of social grooming in primates," *Folia Primatology*, **57**, 121–31.

Dunbar, R I M (1992a), "Time: A hidden constraint on the behavioural ecology of baboons," *Behavioural Ecology and Sociobiology*, **31**, 35–49.

Dunbar, R I M (1992b), "Neocortex size as a constraint on group size in primates," *Journal of Human Evolution*, **20**, 469–93.

Dunbar, R I M (1998), "The social brain hypothesis," *Evolutionary Anthropology*, **6**, 178–90.

Dunbar, R I M (2003), "The social brain: Mind, language, and society in evolutionary perspective," *Annual Review of Anthropology*, **32**, 163–81.

Dunbar, R I M and Bever, J (1998), "Neocortex size determines group size in insectivores and carnivores," *Ethology*, **104**, 695–708.

Dyer, F (1991), "Bees acquire route-based memories but not cognitive maps in a familiar landscape," *Animal Behaviour*, **41**, 239–46.

Elliott, J M and Connolly, K J (1984), "A classification of manipulative hand movements," *Developmental Medicine & Child Neurology*, **26**, 283–96.

Emery, N J (2006), "Cognitive ornithology: The evolution of avian intelligence," *Philosophical Transactions of the Royal Society of London B: Biological Sciences*, **361** (1465), 23–43.

Emery, N J and Clayton, N S (2001), "Effects of experience and social context on prospective caching strategies by scrub jays," *Nature*, **414**, 443–6.

Emery, N J and Clayton, N S (2004), "The mentality of crows: Convergent evolution of intelligence in corvids and apes," *Science*, **306**, 1903–7.

Emery, N J and Clayton, N S (2005), "Evolution of the avian brain and intelligence," *Current Biology*, **15** (23), R946–50.

Emery, N J, Seed, A M, von Bayern, A M, and Clayton, N S (2007), "Cognitive adaptations of social bonding in birds," *Philosophical Transactions of the Royal Society of London B: Biological Sciences*, **362** (1480), 489–505.

Feeney, M C, Roberts, W A, and Sherry, D F (2009), "Memory for what, where, and when in the black-capped chickadee (*Poecile atricapillus*)," *Animal Cognition*, **12** (6), 767–77.

Fisher, J and Hinde, R A (1949), "The opening of milk bottles by birds," *British Birds*, **42**, 347–57.

Flombaum, J I and Santos, L R (2005), "Rhesus monkeys attribute perceptions to others," *Current Biology*, **15** (5), 447–52.

Fodor, J A (1983), *The Modularity of Mind* (Cambridge, MA: MIT Press).

Fox, E, Sitompul, A, and van Schaik, C P (1999), "Intelligent tool use in wild Sumatran orangutans," in S T Parker, H L Miles, and R W Mitchell (eds), *The Mentality of Gorillas and Orangutans* (Cambridge: Cambridge University Press), 99–116.

Fragaszy, D, Izar, P, Visalberghi, E, Ottoni, E B, and de Oliveira, M G (2004), "Wild capuchin monkeys (*Cebus libidinosus*) use anvils and stone pounding tools," *American Journal of Primatology*, **64** (4), 359–66.

Franks, N R and Richardson, T (2006), "Teaching in tandem-running ants," *Nature*, **439** (7073), 153.

Frith, C D and Frith, U (2005), "Theory of mind," *Current Biology*, **15** (17), R644–6.

Frith, U and Happé, F (2005), "Autism spectrum disorder," *Current Biology*, **15** (19), R786–90.

Galdikas, B M F and Vasey, P (1992), "Why are orangutans so smart?," in F D Burtin (ed.), *Social Processes and Mental Abilities in Non-human Primates* (Lewiston, NY: Edward Mellon Press).

Galef, B G (1990), "Tradition in animals: Field observations and laboratory analyses," in M Bekoff and D Jamieson (eds), *Interpretations and Explanations in the Study of Behaviour: Comparative Perspectives* (Boulder, Colorado: Westview Press), 74–95.

Galef, B G (1991), "Information centres of Norway rats: Sites for information exchange and information parasitism," *Animal Behaviour*, **41** (2), 295.

Galef, B G (2003), "'Traditional' foraging behaviours of brown and black rats (*Rattus norwegicus* and *Rattus rattus*)," in D M Fragaszy and S Perry (eds), *The Biology of Traditions: Models and Evidence* (Cambridge: Cambridge University Press), 159–86.

Gallese, V, Fadiga, L, Fogassi, L, and Rizzolatti, G (1996), "Action recognition in the premotor cortex," *Brain*, **119**, 593–609.

Gallup, G G (1970), "Chimpanzees: Self-recognition," *Science*, **167**, 86–7.

Gallup, G G (1979), "Self-awareness in primates," *Scientific American*, **67**, 417–21.

Garber, P (1988), "Foraging decisions during nectar feeding by tamarin monkeys (*Saguinus mystax* and *Saguinus fuscicollis, Callitrichidae*, primates) in Amazonian Peru," *Biotropica*, **20**, 100–6.

Gardner, R A and Gardner, B T (1969), "Teaching sign language to a chimpanzee," *Science*, **165**, 664–72.

Gardner, R A, Gardner, B T, and Van Cantfort, T E (1989), *Teaching Sign Language to Chimpanzees* (New York: SUNY Press).

Genty, E, Breuer, T, Hobaiter, C, and Byrne, R W (2009), "Gestural communication of the gorilla (*Gorilla gorilla*): Repertoire, intentionality and possible origins," *Animal Cognition*, 12, 527–46.

Genty, E and Zuberbuehler, K (2014), "Spatial reference in a bonobo gesture," *Current Biology*, 24, 1601–5.

Gibson, K R (1986), "Cognition, brain size and the extraction of embedded food resources," in J G Else and P C Lee (eds), *Primate Ontogeny, Cognitive and Social Behaviour* (Cambridge: Cambridge University Press), 93–105.

Glickman, S E and Sroges, R W (1964), "Curiosity in zoo animals," *Behaviour*, 26, 151–8.

Goodall, J (1986), *The Chimpanzees of Gombe: Patterns of Behavior* (Cambridge, MA: Harvard University Press).

Goodall, J, Bandora, A, Bergmann, E, Busse, C, Matama, H, Mpongo, E, Pierce, A, and Riss, D (1979), "Intercommunity interactions in the chimpanzee population of the Gombe National Park," in D Hamburg and E R McCown (eds), *The Great Apes* (Menlo Park: Benjamin Cummings), 13–54.

Goody, E N (1995), *Social Intelligence and Interaction. Expressions and Implications of the Social Bias in Human Intelligence* (Cambridge: Cambridge University Press).

Guinet, C and Bouvier, J (1995), "Development of intentional stranding hunting techniques in killer whale (*Orcinus orca*) calves at Crozet Archipelago," *Canadian Journal of Zoology*, 73, 27–33.

Hakeem, A Y, Sherwood, C C, Bonar, C J, Butti, C, Hof, P R, and Allman, J M (2009), "Von Economo neurons in the elephant brain," *Anatomical Record*, 292, 242–8.

Hamilton, W (1855), *Discussions on Philosophy and Literature* (New York: Harper & Brothers).

Hamilton, W D (1971), "Geometry for the selfish herd," *Journal of Theoretical Biology*, 31, 295–311.

Harcourt, A (1992), "Coalitions and alliances: Are primates more complex than non-primates?," in A H Harcourt and F B M de Waal (eds), *Coalitions and Alliances in Humans and Other Animals* (Oxford: Oxford University Press), 445–71.

Hare, B, Call, J, Agnetta, B, and Tomasello, M (2000), "Chimpanzees know what conspecifics do and do not see," *Animal Behaviour*, 59, 771–85.

Hare, B, Call, J, and Tomasello, M (2001), "Do chimpanzees know what conspecifics know?," *Animal Behaviour*, 61 (1), 139–51.

Hare, B and Tomasello, M (2004), "Chimpanzees are more skilful in competitive than in cooperative cognitive tasks," *Animal Behaviour*, 68, 571–81.

Healy, S D and Rowe, C (2007), "A critique of comparative studies of brain size," *Proceedings of the Royal Society of London B: Biological Sciences*, 274 (1609), 453–64.

Held, S, Mendl, M, Devereux, C, and Byrne, R W (2001), "Behaviour of domestic pigs in a visual perspective taking task," *Behaviour*, 138, 1337–54.

Helfman, G S and Schultz, E T (1984), "Social transmission of behavioural traditions in a coral reef fish," *Animal Behaviour*, 32, 379–84.

Helsler, N and Fischer, J (2007), "Gestural communication in Barbary macaques (*Macaca sylvanus*): An overview," in J Call and M Tomasello (eds), *The Gestural Communication of Monkeys and Apes* (Mahwah, NJ: Lawrence Erlbaum Associates).

Hemelrijk, C K (1994a), "Reciprocation in apes: From complex cognition to self-structuring," in W C McGrew, L F Marchant, and T Nishida (ed.), *The Great Apes Revisited* (Cabo San Lucas, Baja, Mexico: Cambridge University Press).

Hemelrijk, C K (1994b), "Support for being groomed in long-tailed macaques," *Animal Behaviour*, **48**, 479–81.

Hemelrijk, C K (1997), "Cooperation without genes, games or cognition," in P Husbands and I Harvey (eds), *Fourth European Conference on Artificial Life* (Cambridge, MA: MIT Press), 511–20.

Hemelrijk, C K and Bolhuis, J J (2011), "A minimalist approach to comparative psychology," *Trends in Cognitive Science*, **15** (5), 185–6.

Heyes, C M (1993a), "Imitation, culture, and cognition," *Animal Behaviour*, **46**, 999–1010.

Heyes, C M (1993b), "Anecdotes, training, trapping and triangulating: Do animals attribute mental states?," *Animal Behaviour*, **46**, 177–88.

Heyes, C M (1994), "Reflections on self-recognition in primates," *Animal Behaviour*, **47**, 909–19.

Heyes, C M (1995), "Self-recognition in primates: Further reflections create a hall of mirrors," *Animal Behaviour*, **50** (6), 1533–42.

Heyes, C M (1998), "Theory of mind in non-human primates," *Behavioral and Brain Sciences*, **21**, 101–48.

Heyes, C M and Ray, E D (2000), "What is the significance of imitation in animals?," *Advances in the Study of Behavior*, **29**, 215–45.

Heyes, C M and Saggerson, A (2002), "Testing for imitative and nonimitative social learning in the budgerigar using a two-object/two-action test," *Animal Behaviour*, **64**, 851–9.

Hobaiter, C and Byrne, R W (2011a), "Serial gesturing by wild chimpanzees: Its nature and function for communication," *Animal Cognition*, **14**, 827–38.

Hobaiter, C and Byrne, R W (2011b), "The gestural repertoire of the wild chimpanzee," *Animal Cognition*, **14**, 745–67.

Hobaiter, C and Byrne, R W (2012), "Gesture use in consortship: Wild chimpanzees' use of gesture for an 'evolutionary urgent' purpose," in S Pika and K Liebal (eds), *Developments in Primate Gesture Research* (Amsterdam: John Benjamins Publishing Company), 129–46.

Hobaiter, C, Poisot, T, Zuberbuehler, K, Hoppit, W J E, and Gruber, T (2014), "Social network analysis shows direct evidence for social transmission of tool use in wild chimpanzees," *PLoS ONE*, **12** (9), e1001960.

Hohmann, G and Fruth, B (2003), "Culture in bonobos? Between-species and within-species variation in behavior," *Current Anthropology*, **44**, 563–71.

Hoppitt, W, Blackburn, L, and Laland, K N (2007), "Response facilitation in the domestic fowl," *Animal Behaviour*, **73** (2), 229–38.

Hoppitt, W and Laland, K N (2008), "Social processes influencing learning in animals: A review of the evidence," *Advances in the Study of Behavior*, **38**, 105–65.

Humle, T and Matsuzawa, T (2002), "Ant dipping among the chimpanzees of Bossou, Guinea, and some comparisons with other sites," *American Journal of Physical Anthropology*, **58**, 133–48.

Humle, T and Snowdon, C T (2008), "Socially biased learning in the acquisition of a complex foraging task in juvenile cotton-top tamarins (*Saguinus oedipus*)," *Animal Behaviour*, **75**, 267–77.

Humphrey, N K (1972), "'Interest' and 'pleasure': Two determinants of a monkey's visual preferences," *Perception*, **1**, 395–416.

Humphrey, N K (1976), "The social function of intellect," in P P G Bateson and R A Hinde (eds), *Growing Points in Ethology* (Cambridge: Cambridge University Press), 303–17.

Jabbi, M., Swart, M., and Keysers, C. (2007), "Empathy for positive and negative emotions in the gustatory cortex," *NeuroImage*, **34** (4), 1744–53.

Janik, V M and Slater, P J B (1997), "Vocal learning in mammals," *Advances in the Study of Behavior*, **26**, 59–99.

Janmaat, K R L, Ban, S D, and Boesch, C (2013), "Chimpanzees use long-term spatial memory to monitor large fruit trees and remember feeding experiences across seasons," *Animal Behaviour*, **86**, 1183–1205.

Janmaat, K R L, Byrne, R W, and Zuberbühler, K (2006), "Primates take weather into account when searching for fruits," *Current Biology*, **16**, 1232–7.

Janson, C H (2000), "Spatial movement strategies: Theory, evidence, and challenges," in S Boinski and P A Garber (eds), *On the Move. How and Why Animals Travel in Groups* (Chicago: Chicago University Press), 165–203.

Janson, C H (2007), "Experimental evidence for route integration and strategic planning in wild capuchin monkeys," *Animal Cognition*, **10** (3), 341–56.

Jarvis, E D and Consortium, A B N (2005), "Avian brains and a new understanding of vertebrate brain evolution," *National Review of Neuroscience*, **6**, 151–9.

Jerison, H J (1963), "Interpreting the evolution of the brain," *Human Biology*, **35**, 263–91.

Jerison, H J (1973), *Evolution of the Brain and Intelligence* (New York: Academic Press).

Jolly, A (1966), "Lemur social behaviour and primate intelligence," *Science*, **153**, 501–6.

Joly, M and Zimmermann, E (2011), "Do solitary foraging nocturnal mammals plan their routes?," *Biology Letters*, **7** (4), 638–40.

Judge, P (1982), "Redirection of aggression based on kinship in a captive group of pigtail macaques (Abstract)," *International Journal of Primatology*, **3**, 301.

Kaminski, J, Call, J, and Tomasello, M (2004), "Body orientation and face orientation: Two factors controlling apes' behavior from humans," *Animal Cognition*, **7**, 216–33.

Kaminski, J, Call, J, and Tomasello, M (2008), "Chimpanzees know what others know, but not what they believe," *Cognition*, **109** (2), 224–34.

Kaminski, J, Riedel, J, Call, J, and Tomasello, M (2005), "Domestic goats, *Capra hircus*, follow gaze direction and use social cues in an object choice task," *Animal Behaviour*, **69** (1), 11–18.

Kandel, E R (1979), *Behavioral Biology of Aplysia* (San Francisco: W H Freeman).

Karin-D'Arcy, M and Povinelli, D J (2002), "Do chimpanzees know what each other see? A closer look," *International Journal of Comparative Psychology*, **15**, 21–54.

Karmiloff-Smith, A (1993), *Beyond Modularity: A Developmental Perspective on Cognitive Science* (Cambridge, MA: Bradford/MIT Press).

Keysers, C (2011), *The Empathic Brain. How the Discovery of Mirror Neurons Changes our Understanding of Human Nature* (Social Brain Press).

King, B J (1986), "Extractive foraging and the evolution of primate intelligence," *Human Evolution*, **1** (4), 361–72.

King, B J (2004), *The Dynamic Dance. Nonvocal Communication in African Great Apes* (Cambridge, MA: Harvard University Press).

Kitchen, D M, Cheney, D L, and Seyfarth, R M (2005), "Male chacma baboons (Papio hamadryas ursinus) discriminate loud call contests between rivals of different relative ranks." Animal Cognition 8(1), 1–6.

Kobayashi, H and Kohshima, S (1997), "Unique morphology of the human eye," Nature, 387, 767–8.

Koops, K, Visalberghi, E, and van Schaik, C P (2014), "The ecology of primate material culture," Biology Letters, 10 (11), 20140508.

Krebs, J R and Dawkins, R (1984), "Animal signals: Mind reading and manipulation," in J R Krebs and N B Davies (eds), Behavioural Ecology: An Evolutionary Approach (Oxford: Blackwell), 380–401.

Kuczaj, S, Tranel, K, Trone, M, and Hill, H (2001), "Are animals capable of deception or empathy? Implications for animal consciousness and animal welfare," Animal Welfare, 10, S161–73.

Kummer, H (1967), "Tripartite relations in hamadryas baboons," in S A Altmann (ed.), Social Communication among Primates (Chicago: University of Chicago Press), 63–71.

Kummer, H (1982), "Social knowledge in free-ranging primates," in D R Griffin (ed.), Animal Mind—Human Mind (New York: Springer-Verlag).

Laland, K N (1996), "Is social learning always locally adaptive?," Animal Behaviour, 52 (3), 637–40.

Laland, K N, Atton, N, and Webster, M M (2011), "From fish to fashion: Experimental and theoretical insights into the evolution of culture," Philosophical Transactions of the Royal Society of London B: Biological Sciences, 366 (1567), 958–68.

Laland, K N and Hoppit, W J E (2003), "Do animals have culture?," Evolutionary Anthropology, 12, 150–9.

Lefebvre, L (2005), "Ecology and evolution of social learning," International Conference on Social Learning (St Andrews, Scotland).

Lefebvre, L, Reader, S M, and Sol, D (2004), "Brains, innovations and evolution," Brain, Behavior and Evolution, 63, 233–46.

Lefebvre, L, Whittle, P, Lascaris, E, and Finkelstein, A (1997), "Feeding innovations and forebrain size in birds," Animal Behaviour, 53, 549–60.

Lehmann, H E (1979), "Yawning: A homeostatic reflex and its psychological significance," Bulletin of the Menninger Clinic, 43, 123–36.

Leland, S (1997), Peaceful Kingdom. Random Acts of Kindness by Animals (California, USA: Conari Press).

Lenneberg, E H (1968), The Biological Basis for Language (New York: Wiley).

Liebal, K (2007), "Gestures in orangutans," in J Call and M Tomasello (eds), The Gestural Communication of Apes and Monkeys (Mahwah, NJ: Lawrence Erlbaum Associates), 69–98.

Liebal, K, Call, J, and Tomasello, M (2004a), "Use of gesture sequences in chimpanzees," American Journal of Primatology, 64 (4), 377–96.

Liebal, K, Pika, S, Call, J, and Tomasello, M (2004b), "To move or not to move: How apes adjust to the attentional state of others," Interaction Studies, 5, 199–219.

Liebal, K, Pika, S, and Tomasello, M (2006), "Gestural communication of orangutans (Pongo pygmaeus)," Gesture, 6, 1–38.

Limongelli, L, Boysen, S T, and Visalberghi, E (1995), "Comprehension of cause–effect relations in a tool-using task by chimpanzees (Pan troglodytes)," Journal of Comparative Psychology, 109, 18–26.

Loretto, M C, Schloegl, C, and Bugnyar, T (2010), "Northern bald ibises follow others' gaze into distant space but not behind barriers," *Biology Letters*, 6 (1), 14–17.
Loucks, J and Baldwin, D (2009), "Sources of information for discriminating dynamic human actions," *Cognition*, 111, 84–97.
Machiavelli, N (1532/1979), *The Prince* (Harmondsworth, Middlesex: Penguin Books).
Mackinnon, J (1978), *The Ape within Us* (London: Collins).
MacLeod, C E, Zilles, K, Schleicher, A, Rilling, J K, and Gibson, K R (2003), "Expansion of the neocerebellum in Hominoidea," *Journal of Human Evolution*, 44 (4), 401–29.
Macphail, E M (1982), *Brain and Intelligence in Vertebrates* (Oxford: Clarendon Press).
Macphail, E M (1985), "Vertebrate intelligence: The null hypothesis," in L Weiskrantz (ed.), *Animal Intelligence* (Oxford: Clarendon Press), 37–50.
Marr, D (1969), "A theory of cerebellar cortex," *Journal of Physiology*, 202, 437–70.
Mason, W A and Hollis, J H (1962), "Communication between young rhesus monkeys," *Animal Behaviour*, 10, 211–21.
Matsuzawa, T (2001), "Primate foundations of human intelligence: A view of tool use in nonhuman primates and fossil hominids," in T Matsuzawa (ed.), *Primate Origins of Human Cognition and Behavior* (Tokyo: Springer-Verlag), 3–25.
Matsuzawa, T and Yamakoshi, G (eds) (1996), "Comparisons of chimpanzee material culture between Bossou and Nimba, West Africa," in A E Russon, K A Bard, and S T Parker (eds), *Reaching into Thought: The Minds of Great Apes* (Cambridge: Cambridge University Press), 211–34.
McComb, K, Baker, L, and Moss, C (2006), "African elephants show high levels of interest in the skulls and ivory of their own species," *Biology Letters*, 2 (1), 26–8.
McComb, K, Shannon, G, Sayialel, K N, and Moss, C (2014), "Elephants can determine ethnicity, gender, and age from acoustic cues in human voices," *Proceedings of the National Academy of Sciences*, 111, 5433–5438.
McGrew, W C (1989), "Why is ape tool use so confusing?," in V Standen and R A Foley (eds), *Comparative Socioecology: The Behavioural Ecology of Humans and Other Mammals* (Oxford: Blackwell Scientific Publications), 457–72.
McGrew, W C (1992), *Chimpanzee Material Culture: Implications for Human Evolution* (Cambridge: Cambridge University Press).
McGrew, W C and Tutin, C E G (1978), "Evidence for a social custom in wild chimpanzees?," *Man*, 13, 234–51.
McKiggan, H (1995), "Cognitive capacities underlying the use of mirror and video images by two species of mangabey (*Cercocebus t. torquatus* and *C. a. albigena*)," Ph.D. (University of St Andrews).
Melis, A P, Call, J, and Tomasello, M (2006a), "Chimpanzees (*Pan troglodytes*) conceal visual and auditory information from others," *Journal of Comparative Psychology*, 120 (2), 154–62.
Melis, A P, Hare, B, and Tomasello, M (2006b), "Chimpanzees recruit the best collaborators," *Science*, 311 (5765), 1297–300.
Mercader, J, Barton, H, Gillespie, J, Harris, J, Kuhn, S, Tyler, R, and Boesch, C (2007), "4,300-year-old chimpanzee sites and the origins of percussive stone technology," *Proceedings of the National Academy of Sciences*, 104, 1–7.
Miles, H L (1986), "Cognitive development in a signing orangutan," *Primate Report*, 14, 179–80.

Mitani, J C and Watts, D P (2001), "Why do chimpanzees hunt and share meat?," *Animal Behaviour*, **61**, 915–24.

Mitchell, R W and Hamm, M (1997), "The interpretation of animal psychology: Anthropomorphism or behavior reading?," *Behaviour*, **134**, 173–204.

Morris, R G (1981), "Spatial localization does not require the presence of local cues," *Learning and Motivation*, **12**, 239–60.

Morton, J and Frith, U (2004), *Understanding Developmental Disorders: A Causal Modeling Approach* (Oxford: Blackwell Publishers).

Moura, A C de A and Lee, P C (2004), "Capuchin stone tool use in Caatinga dry forest," *Science (Washington, DC)*, **306** (5703), 1909.

Napier, J R (1961), "Prehensility and opposability in the hands of primates," *Symposia of the Zoological Society of London*, **5**, 115–32.

Newell, A and Simon, H A (1972), *Human Problem Solving* (New York: Prentice-Hall).

Nisbett, R E and Ross, L (1980), *Human Inference: Strategies and Shortcomings of Social Judgement* (Englewood Cliffs, NJ: Prentice-Hall).

Noser, R and Byrne, R W (2007a), "Mental maps in chacma baboons (*Papio ursinus*): Using intergroup encounters as a natural experiment," *Animal Cognition*, **10**, 331–40.

Noser, R and Byrne, R W (2007b), "Travel routes and planning of visits to out-of-sight resources in wild chacma baboons, *Papio ursinus*," *Animal Behaviour*, **73**, 257–66.

Noser, R and Byrne, R W (2015), "Wild chacma baboons (*Papio ursinus*) remember single foraging episodes," *Animal Cognition*, **18**, 921–9.

Onishi, K H and Baillargeon, R (2005), "Do 15-month-old infants understand false beliefs?," *Science*, **308**, 255–8.

Parker, S T (2015), "Re-evaluating the extractive foraging hypothesis," *New Ideas in Psychology*, **37**, 1–12.

Parker, S T and Gibson, K R (1977), "Object manipulation, tool use, and sensorimotor intelligence as feeding adaptations in cebus monkeys and great apes," *Journal of Human Evolution*, **6**, 623–41.

Parker, S T and Gibson, K R (1979), "A developmental model for the evolution of language and intelligence in early hominids," *Behavioural and Brain Sciences*, **2**, 367–408.

Parr, L A, Waller, B M, and Fugate, J (2005), "Emotional communication in primates: Implications for neurobiology," *Current Opinion in Neurobiology*, **15** (6), 716–20.

Patterson, F and Linden, E (1981), *The Education of Koko* (New York: Holt, Rinehart, and Linden).

Paz-y-Mino, G, Bond, A B, Kamil, A C, and Balda, R P (2004), "Pinyon jays use transitive inference to predict social dominance," *Nature*, **430**, 778–81.

Pepperberg, I M (1999), *The Alex Studies. Cognitive and Communicative Abilities of Grey Parrots* (Cambridge, MA: Harvard University Press).

Pepperberg, I M and Gordon, J D (2005), "Number comprehension by a grey parrot (*Psittacus erithacus*), including a zero-like concept," *Journal of Comparative Psychology*, **119** (2), 197–209.

Perry, S, Baker, M, Fedigan, L, Gros-Louis, J, Jack, K, MacKinnon, K C, Manson, J H, Panger, M, Pyle, K, and Rose, L (2003), "Social conventions in wild white-faced capuchin monkeys—evidence for traditions in a neotropical primate," *Current Anthropology*, **44**, 241–68.

Pfungst, O (1911), *Clever Hans. The Horse of Mr. von Osten: A Contribution to Experimental Animal and Human Psychology* (trans. C L Rahn; originally published in German, 1907) (New York: Henry Holt).

Pika, S (2007a), "Gestures in subadult gorillas (*Gorilla gorilla*)," in J Call and M Tomasello (eds), *The Gestural Communication of Apes and Monkeys* (Mahwah, NJ: Lawrence Erlbaum Associates), 99–130.

Pika, S (2007b), "Gestures in subadult bonobos (*Pan paniscus*)," in J Call and M Tomasello (eds), *The Gestural Communication of Apes and Monkeys* (Mahwah, NJ: Lawrence Erlbaum Associates), 41–67.

Pilbeam, D and Smith, R (1981), "New skull remains of *Sivapithecus* from Pakistan," *Memoirs of the Geological Survey of Pakistan*, **11**, 1–13.

Pinker, S (1994), *The Language Instinct. The New Science of Language and Mind* (Harmondsworth: Penguin).

Plooij, F X (1984), *The Behavioral Development of Free-living Chimpanzee Babies and Infants* (Norwood, NJ: Ablex Publishing Corporation).

Plotnik, J M, de Waal, F B M, and Reiss, D (2006), "Self-recognition in an Asian elephant," *Proceedings of the National Academy of Sciences*, **103**, 17053–17057.

Plotnik, J M, Lair, R, Suphachoksahakun, W, and de Waal, F B M (2011), "Elephants know when they need a helping trunk in a cooperative task," *Proceedings of the National Academy of Sciences*, **108** (12), 5116–21.

Polansky, L, Kilian, W, and Wittemeyer, G (2015), "Elucidating the significance of spatial memory on movement decisions by African savannah elephants using state–space models," *Proceedings of the Royal Society of London B*, **282**, 20143042.

Poole, J H and Granli, P K (2011), "Signals, gestures, and behavior of African elephants," in C J Moss, H J Croze, and P C Lee (eds), *The Amboseli Elephants: A Long-Term Perspective on a Long-Lived Mammal* (Chicago: University of Chicago Press), 109–24.

Povinelli, D J, Bering, J M, and Giambrone, S (2000), "Towards a science of other minds: Escaping the argument by analogy," *Cognitive Science*, **24** (3), 509–41.

Povinelli, D J and Cant, J G H (1995), "Arboreal clambering and the evolution of self-conception," *Quarterly Journal of Biology*, **70**, 393–421.

Povinelli, D J and Eddy, T J (1996), "What young chimpanzees know about seeing," *Monographs of the Society for Research in Child Development*, **61** (3), 1–189.

Povinelli, D J, Nelson, K E, and Boysen, S T (1990), "Inferences about guessing and knowing by chimpanzees (*Pan troglodytes*)," *Journal of Comparative Psychology*, **104**, 203–10.

Povinelli, D J, Nelson, K E, and Boysen, S T (1992), "Comprehension of role reversal in chimpanzees: Evidence of empathy?," *Animal Behaviour*, **43**, 633–40.

Povinelli, D J and O'Neill, D K (2000), "Do chimpanzees use gestures to instruct each other during cooperative situations?," in S Baron-Cohen, H Tager-Flusberg, and D J Cohen (eds), *Understanding Other Minds: Perspectives from Autism* (2nd edn) (Oxford: Oxford University Press), 459–87.

Povinelli, D J, Parks, K A, and Novak, M A (1991), "Do rhesus monkeys (*Macaca mulatta*) attribute knowledge and ignorance to others?," *Journal of Comparative Psychology*, **105**, 318–25.

Povinelli, D J, Parks, K A, and Novak, M A (1992b), "Role reversal by rhesus monkeys, but no evidence of empathy," *Animal Behaviour*, **44**, 269–81.

Povinelli, D J, Rulf, A B, and Bierschwale, D T (1994), "Absence of knowledge attribution and self-recognition in young chimpanzees (*Pan troglodytes*)," *Journal of Comparative Psychology*, **108**, 74–80.

Povinelli, D J and Vonk, J (2003), "Chimpanzee minds: Suspiciously human?," *Trends in Cognitive Sciences*, **7** (4), 157–60.

Premack, D and Woodruff, G (1978), "Does the chimpanzee have a theory of mind?," *Behavioural and Brain Sciences*, **1**, 515–26.

Prior, H, Schwarz, A., and Guentuerkuen, O. (2008), "'Mirror-induced behavior in the magpie (*Pica pica*): Evidence of self-recognition,'" *PLoS Biology*, **6**, 1642–50.

Raihani, N J and Ridley, A R (2008), "Experimental evidence for teaching in wild pied babblers," *Animal Behaviour*, **75** (1), 3–11.

Rapaport, L G and Brown, G R (2008), "Social influences on foraging behavior in young nonhuman primates: Learning what, where, and how to eat," *Evolutionary Anthropology: Issues, News, and Reviews*, **17** (4), 189–201.

Reader, S M, Hager, Y, and Laland, K N (2011), "The evolution of primate general and cultural intelligence," *Philosophical Transactions of the Royal Society of London B: Biological Sciences*, **366** (1567), 1017–27.

Reader, S M and Laland, K N (2001), "Social intelligence, innovation and enhanced brain size in primates," *Proceedings of the National Academy of Sciences*, **99**, 4436–41.

Reid, P J (2009), "Adapting to the human world: Dogs' responsiveness to our social cues," *Behavioural Processes*, **80** (3), 325–33.

Reiss, D and Marino, L (2001), "Mirror self-recognition in the bottlenose dolphin: A case of cognitive convergence," *Proceedings of the National Academy of Sciences*, **98** (10), 5937–42.

Rendell, L and Whitehead, H (2001), "Culture in whales and dolphins," *Behavioral and Brain Sciences*, **24**, 309–82.

Riedman, M L, Staedier, M M, Estes, J A, and Hrabrich, B (1989), "The transmission of individually distinctive foraging strategies from mother to offspring in sea otters (*Enhydra lutris*)," Eighth Biennial Conference on the Biology of Marine Mammals (Pacific Grove, CA).

Ristau, C (1991), "Aspects of the cognitive ethology of an injury-feigning bird, the piping plover," in C Ristau (ed), *Cognitive ethology: the minds of other animals*. (Hillsdale, NJ: Lawrence Erlbaum Associates), 91–126.

Rizzolatti, G, Fadiga, L, Fogassi, L, and Gallese, V (1996), "Premotor cortex and the recognition of motor actions," *Brain Research*, **3**, 131–41.

Rizzolatti, G, Fadiga, L, Fogassi, L, and Gallese, V (2002), "From mirror neurons to imitation: Facts and speculations," in A N Meltzoff and W Prinz (eds), *The Imitative Mind: Development, Evolution, and Brain Bases* (Cambridge: Cambridge University Press), 247–66.

Rosch, E, Mervis, C B, Gray, W D, Johnson, D M, and Boyes-Braem, P (1976), "Basic objects in natural categories," *Cognitive Psychology*, **8**, 382–439.

Ruiz, A M, Gómez, J C, Roeder, J J, and Byrne, R W (2009), "Gaze following and gaze priming in lemurs," *Animal Cognition*, **12**, 427–34.

Ruiz, A M, Marticorena, D C, Mukerji, C, Goddu, A, and Santos, L R (2010), "Do rhesus monkeys reason about false beliefs?," International Primatological Society 23rd Congress (Kyoto).

Russon, A E (1998), "The nature and evolution of intelligence in orangutans (*Pongo pygmaeus*)," *Primates*, **39** (4), 485–503.

Russon, A E and Andrews, K (2011), "Orangutan pantomime: Elaborating the message," *Biology Letters*, **7** (4), 627–30.

Sabbatini, G, Truppa, V, Hribar, A, Gambetta, B, Call, J, and Visalberghi, E (2012), "Understanding the functional properties of tools: Chimpanzees (*Pan troglodytes*) and capuchin monkeys (*Cebus apella*) attend to tool features differently," *Animal Cognition*, **15** (4), 577–90.

Saffran, J R, Aslin, R N, and Newport, E L (1996), "Statistical learning by 8-month-old infants," *Science*, **274**, 1926–8.

Sambrook, T and Whiten, A (1997), "On the nature of complexity in cognitive and behavioural science," *Theory & Psychology*, **7**, 191–213.

Santos, L R, Marticorena, D, and Goddu, A (2007), "Do monkeys reason about the false beliefs of others?," 14th Biennial Meeting of the Society for Research in Child Development (Boston, MA).

Santos, L R, Nissen, A G, and Ferrugia, J A (2006), "Rhesus monkeys, *Macaca mulatta*, know what others can and cannot hear," *Animal Behaviour*, **71** (5), 1175–81.

Sanz, C, Call, J, and Morgan, D (2009), "Design complexity in termite-fishing tools of chimpanzees (*Pan troglodytes*)," *Biology Letters*, **5** (3), 293–6.

Sanz, C and Morgan, D (2009), "Complexity of chimpanzee tool using behaviors," in E V Lonsdorf, S R Ross, and T Matsuzawa (eds), *The Mind of the Chimpanzee: Ecological and Experimental Perspectives* (Chicago: University of Chicago Press), 127–40.

Savage-Rumbaugh, E S, Murphy, J, Sevcik, R A, Brakke, K E, Williams, S L, and Rumbaugh, D M (1993), "Language comprehension in ape and child," *Monographs of the Society for Research in Child Development*, **58** (3–4), 1–222.

Schel, A M, Machanda, Z, Townsend, S W, Zuberbühler, K, and Slocombe, K E (2013), "Chimpanzee food calls are directed at specific individuals," *Animal Behaviour*, **86** (5), 955–65.

Seed, A M, Call, J, Emery, N J, and Clayton, N S (2009), "Chimpanzees solve the trap problem when the confound of tool-use is removed," *Journal of Experimental Psychology: Animal Behaviour Processes*, **35** (1), 23–34.

Seed, A M, Clayton, N S, and Emery, N J (2007), "Postconflict third-party affiliation in rooks, *Corvus frugilegus*," *Current Biology*, **17** (2), 152–8.

Seyfarth, R M and Cheney, D L (1984), "Grooming, alliances and reciprocal altruism in vervet monkeys," *Nature*, **308**, 541–2.

Seyfarth, R M and Cheney, D L (2002), "What are big brains for?," *Proceedings of the National Academy of Sciences*, **99**, 4141–2.

Seyfarth, R M, Cheney, D L, and Marler, P (1980a), "Vervet monkey alarm calls: Semantic communication in a free-ranging primate," *Animal Behaviour*, **28**, 1070–94.

Seyfarth, R M, Cheney, D L, and Marler, P (1980b), "Monkey responses to three different alarm calls: Evidence of predator classification and semantic communication," *Science*, **210**, 801–3.

Sharman, M (1981), "Feeding, ranging and social organisation of the Guinea baboon," Ph.D. (St Andrews).

Shepherd, S V and Platt, M L (2008), "Spontaneous social orienting and gaze following in ringtailed lemurs (*Lemur catta*)," *Animal Cognition*, **11** (1), 13–20.

Shettleworth, S J (1998), *Cognition, Evolution and Behavior* (New York: Oxford University Press).

Shettleworth, S J (2010), "Clever animals and killjoy explanations in comparative psychology," *Trends in Cognitive Science*, **14** (11), 477–81.

Shultz, S and Dunbar, R I M (2006), "Both social and ecological factors predict ungulate brain size," *Proceedings of the Royal Society of London B: Biological Sciences*, **273** (1583), 207–15.

Shumaker, R, Walkup, K R, and Beck, B (2011), *Animal Tool Behavior: The Use and Manufacture of Tools by Animals* (Baltimore, Maryland: The Johns Hopkins University Press).

Sigg, H and Stolba, A (1981), "Home range and daily march in a hamadryas baboon troop," *Folia Primatologica*, **36**, 40–75.

Smet, A F and Byrne, R W (2014), "African elephants (*Loxodonta africana*) recognize visual attention from face and body orientation," *Biology Letters*, **10**, 20140428.

Southgate, V, Senju, A, and Csibra, G (2007), "Action anticipation through attribution of false-belief by 2-year-olds," *Psychological Science*, **18**, 587–92.

Sternberg, R J (1985), "General intellectual ability," in R J Sternberg (ed.), *Human Abilities. An Information Processing Account* (New York: W H Freeman).

Stokes, E J and Byrne, R W (2001), "Cognitive capacities for behavioural flexibility in wild chimpanzees (*Pan troglodytes*): The effect of snare injury on complex manual food processing," *Animal Cognition*, **4**, 11–28.

Stokes, E J, Quiatt, D, and Reynolds, V (1999), "Snare injuries to chimpanzees (*Pan troglodytes*) at 10 study sites in East and West Africa," *American Journal of Primatology*, **49**, 104–5.

Struhsaker, T T (1967), "Behaviour of vervet monkeys," *University of California Publications of Zoology*, **82**, 1–74.

Suzuki, S, Kuroda, S, and Nishihara, T (1995), "Tool-set for termite-fishing by chimpanzees in the Ndoki Forest, Congo," *Behaviour*, **132**, 219–35.

Takasaki, H (1983), "Mahale chimpanzees taste mangoes—toward acquisition of a new food item?," *Primates*, **24** (2), 273–5.

Tanner, J E and Byrne, R W (1993), "Concealing facial evidence of mood: Evidence for perspective-taking in a captive gorilla?," *Primates*, **34**, 451–6.

Tanner, J E and Byrne, R W (1996), "Representation of action through iconic gesture in a captive lowland gorilla," *Current Anthropology*, **37**, 162–73.

Tanner, J E and Byrne, R W (1999), "The development of spontaneous gestural communication in a group of zoo-living lowland gorillas," in S T Parker, R W Mitchell, and H L Miles (eds), *The Mentalities of Gorillas and Orangutans. Comparative Perspectives* (Cambridge: Cambridge University Press), 211–39.

Taylor, R J, Balph, D F, and Balph, M H (1990), "The evolution of alarm calling: A cost-benefit analysis," *Animal Behaviour*, **39** (5), 860.

Tebbich, S, Taborsky, M, Fessl, B, and Blomqvist, D (2001), "Do woodpecker finches acquire tool-use by social learning?," *Proceedings of the Royal Society of London B: Biological Sciences*, **268** (1482), 2189–93.

Tennie, C, Call, J, and Tomasello, M (2009), "Ratcheting up the ratchet: On the evolution of cumulative culture," *Philosophical Transactions of the Royal Society of London B: Biological Sciences*, **364** (1528), 2405–15.

Terborgh, J, Robinson, S K, Parker, T A, Munn, C A, and Pierpont, N (1990), "Structure and organization of an Amazonian forest bird community," *Ecological Monographs*, 60, 213–38.

Thompson, E (2001), "Empathy and consciousness," *Journal of Consciousness Studies*, 8 (5–7), 1–32.

Thorndike, E L (1898), "Animal intelligence: An experimental study of the associative process in animals," *Psychological Review and Monograph*, 2 (8), 551–3.

Thornton, A and McAuliffe, K (2006), "Teaching in wild meerkats," *Science*, 313 (5784), 227–9.

Thornton, A and Raihani, N J (2010), "Identifying teaching in wild animals," *Learning and Behaviour*, 38 (3), 297–309.

Tomasello, M and Call, J (1997), *Primate Cognition* (New York: Oxford University Press).

Tomasello, M, Call, J, and Hare, B (1998), "Five primate species follow the visual gaze of conspecifics," *Animal Behaviour*, 55 (4), 1063–9.

Tomasello, M, Call, J, and Hare, B (2003), "Chimpanzees understand psychological states—the question is which ones and to what extent," *Trends in Cognitive Sciences*, 7 (4), 153–6.

Tomasello, M, Call, J, Nagell, C, Olguin, R, and Carpenter, M (1994), "The learning and use of gestural signals by young chimpanzees: A trans-generational study," *Primates*, 35, 137–54.

Tomasello, M, George, B, Kruger, A, Farrar, J, and Evans, E (1985), "The development of gestural communication in young chimpanzees," *Journal of Human Evolution*, 14, 175–86.

Tomasello, M, Gust, D, and Frost, T A (1989), "A longitudinal investigation of gestural communication in young chimpanzees," *Primates*, 30, 35–50.

Tomasello, M, Hare, B, and Agnetta, B (1999), "Chimpanzees, *Pan troglodytes*, follow gaze direction geometrically," *Animal Behaviour*, 58, 769–77.

Tomasello, M, Kruger, A C, and Ratner, H H (1993), "Cultural learning," *Behavioral and Brain Sciences*, 16, 495–552.

Tulving, E (1972), "Episodic and semantic memory," in E Tulving and W Donaldson (eds), *Organization of Memory* (New York: Academic Press), 381–403.

Tulving, E (2002), "Episodic memory: From mind to brain," *Annual Review of Psychology*, 53, 1–25.

Udell, M A R, Dorey, N R, and Wynne, C D L (2008), "Wolves outperform dogs in following human social cues," *Animal Behaviour*, 76 (6), 1767–73.

van Schaik, C P (1983), "Why are diurnal primates living in groups?," *Behaviour*, 87, 120–47.

van Schaik, C P, Ancrenaz, M, Borgen, G, Galdikas, B M F, Knott, C D, Singleton, I, Suzuki, A, Utami, S S, and Merrill, M (2003), "Orangutan cultures and the evolution of material culture," *Science*, 299, 102–5.

van Schaik, C P, Fox, E A, and Sitompul, A F (1996), "Manufacture and use of tools in wild Sumatran orangutans. Implications for human evolution," *Naturwissenschaften*, 83, 186–8.

Viranyi, Z, Gacsi, M, Kubinyi, E, Topal, J, Belenyi, B, Ujfalussy, D, and Miklosi, A (2008), "Comprehension of human pointing gestures in young human-reared wolves (*Canis lupus*) and dogs (*Canis familiaris*)," *Animal Cognition*, 11 (3), 373–87.

Visalberghi, E and Limongelli, L (1994), "Lack of comprehension of cause–effect relationships in tool-using capuchin monkeys (*Cebus apella*)," *Journal of Comparative Psychology*, **103**, 15–20.

Visalberghi, E, Spagnoletti, N, Ramos da Silva, E D, Andrade, F R, Ottoni, E, Izar, P, and Fragaszy, D (2009), "Distribution of potential suitable hammers and transport of hammer tools and nuts by wild capuchin monkeys," *Primates*, **50** (2), 95–104.

Voronov, L N, Bogoslovskaya, L G, and Markova, E G (1994), "A comparative study of the morphology of forebrain in corvidae in view of their trophic specialization (in Russian)," *Zoologičeskij žurnal.*, **73**, 82–96.

Wallman, J (1990), *Aping Language* (Cambridge: Cambridge University Press).

Warneken, F, Hare, B, Melis, A P, Hanus, D, and Tomasello, M (2007), "Spontaneous altruism by chimpanzees and young children," *PLoS Biology*, **5**, e184.

Warneken, F and Tomasello, M (2006), "Altruistic helping in human infants and young chimpanzees," *Science*, **311**, 1301–3.

Warner, R R (1988), "Traditionality of mating-site preferences in a coral reef fish," *Nature*, **335**, 719–21.

Warren, J M (1973), "Learning in vertebrates," in D A Dewsbury and D A Rethlingshafer (eds), *Comparative Psychology: A Modern Survey* (New York: McGraw Hill), 471–509.

Watts, D P and Mitani, J C (2001), "Boundary patrols and intergroup encounters in wild chimpanzees," *Behaviour*, **138**, 299–327.

Watve, M, Thakar, J, Kale, A, Puntambekar, S, Shaikh, I, Vaze, K, Jog, M, and Paranjape, S (2002), "Bee-eaters (*Merops orientalis*) respond to what a predator can see," *Animal Cognition*, **5** (4), 253–9.

Whiten, A (1998), "Imitation of the sequential structure of actions by chimpanzees (*Pan troglodytes*)," *Journal of Comparative Psychology*, **112**, 270–81.

Whiten, A and Byrne, R W (1988a), "The manipulation of attention in primate tactical deception," in R W Byrne and A Whiten (eds), *Machiavellian Intelligence: Social Expertise and the Evolution of Intellect in Monkeys, Apes and Humans* (Oxford: Clarendon Press), 211–23.

Whiten, A and Byrne, R W (1988b), "Tactical deception in primates," *Behavioral and Brain Sciences*, **11**, 233–73.

Whiten, A and Byrne, R W (1991), "The emergence of metapresentation in human ontogeny and primate phylogeny," in A Whiten (ed.), *Natural Theories of Mind: Evolution, Development and Simulation of Everyday Mindreading* (Oxford: Basil Blackwell), 267–81.

Whiten, A, Goodall, J, McGrew, W C, Nishida, T, Reynolds, V, Sugiyama, Y, Tutin, C E G, Wrangham, R W, and Boesch, C (1999), "Cultures in chimpanzees," *Nature*, **399**, 682–5.

Whiten, A, Goodall, J, McGrew, W C, Nishida, T, Reynolds, V, Sugiyama, Y, Tutin, C E G, Wrangham, R W, and Boesch, C (2001), "Charting cultural variation in chimpanzees," *Behaviour*, **138**, 1481–516.

Whiten, A, Spiteri, A, Horner, V, Bonnie, K E, Lambeth, S P, Schapiro, S J, and de Waal, F B M (2007), "Transmission of multiple traditions within and between chimpanzee groups," *Current Biology*, **17** (12), 1038–43.

Wich, S A and de Vries, H (2006), "Male monkeys remember which group members have given alarm calls," *Proceedings of the Royal Society of London B: Biological Sciences*, **273** (1587), 735–40.

Wicker, B, Keysers, C, Plailly, J, Royet, J, Gallese, V, and Rizzolatti, G (2003), "Both of us disgusted in my insula: The common neural basis of seeing and feeling disgust," *Neuron*, **40** (3), 655–64.

Wilcox, S and Jackson, R (2002), "Jumping spider tricksters: Deceit, predation, and cognition," in M Bekoff, C Allen, and G M Burghardt (eds), *The Cognitive Animal. Empirical and Theoretical Perspectives on Animal Cognition* (Cambridge, MA: MIT Press), 27–33.

Wilkinson, A, Mandl, I, Bugnyar, T, and Huber, L (2010), "Gaze following in the red-footed tortoise (*Geochelone carbonaria*)," *Animal Cognition*, **13** (5), 765–9.

Willems, E P and Hill, R A (2009), "Predator-specific landscapes of fear and resource distribution: Effects on spatial range use," *Ecology*, **90** (2), 546–55.

Wrangham, R (2009), *Catching Fire. How Cooking Made us Human* (New York: Basic Books).

Young, R M (1978), "Strategies and structure of a cognitive skill," in G Underwood (ed.), *Strategies of Information Processing* (New York: Academic Press).

Young, R M and O'Shea, T (1981), "Errors in children's subtraction," *Cognitive Science*, **5**, 153–77.

Zacks, J M (2004), "Using movement and intentions to understand simple events," *Cognitive Science*, **28** (6), 979–1008.

Zacks, J M, Kumar, S, Abrams, R A, and Mehta, R (2009), "Using movement and intentions to understand human activity," *Cognition*, **112** (2), 201–16.

Zacks, J M, Tversky, B, and Iyer, G (2001), "Perceiving, remembering, and communicating structure in events," *Journal of Experimental Psychology: General*, **130** (1), 29–58.

Zentall, T R and Akins, C K (1996), "Imitative learning in male Japanese quail (*Coturnix japonica*) using the two-action method," *Journal of Comparative Psychology*, **110**, 316–20.

Zuberbühler, K (2000a), "Causal cognition in a non-human primate: Field playback experiments with Diana monkeys," *Cognition*, **76**, 195–207.

Zuberbühler, K (2000b), "Causal knowledge of predators' behaviour in wild Diana monkeys," *Animal Behaviour*, **59**, 209–20.

Zuckerman, S (1932), *The Social Life of Monkeys and Apes* (London: Kegan, Paul, Trench, Trubner).

Author index

A
Akins, C K 136
Allman, J M 107
Anderson, J R 92, 97, 98
Andrews, K 28
Apperly, I 166
Arnold, K 98
Aureli, F 98

B
Baillargeon, R 88
Balda, R P 117
Baldwin, D 117, 139, 153
Ban, S D 120
Bargh, J A 68, 153, 166
Barton, R A 60, 66, 98, 158
Bates, L A 96, 98, 104, 129
Beck, B B 80
Benhamou, S 122
Bever, J 60
Boesch, C 76, 89, 102, 103, 120, 143
Boesch, H 120, 143
Boesch-Achermann, H 89
Bolhuis, J J 67
Bouvier, J 104
Bovet, D 55
Boysen, S T 11, 20
Brewer, S 111
Brooks, R 40, 50
Brothers, L 59
Brown, G R 103
Brüne, M 61
Bugnyar, T 48, 49
Busse, C D 89
Butterfill, S A 166
Byrne, J M E 81, 113, 143, 145, 152
Byrne, R W 12, 23, 24, 25, 27, 28, 29, 31, 32, 34, 35, 50, 53, 56, 58, 66, 67, 70, 71, 73, 77, 78, 79, 81, 86, 89, 98, 103, 112, 113, 118, 120, 122, 123, 126, 137, 138, 139, 143, 145, 148, 149, 152

C
Caldwell, D K 100
Caldwell, M C 100
Call, J 23, 36, 48, 85, 86, 87, 117, 145, 155
Cant, J G H 93, 94

Caro, T M 100, 101, 102
Cartmill, E A 24, 31, 32
Chance, M R A 56
Chartrand, T L 68, 153, 166
Cheney, D L 15, 17, 18, 54, 55, 61, 62
Christel, M I 142, 164
Clayton, N S 8, 63, 87, 118
Clutton-Brock, J 41
Cochet, H 53
Collins, D A 77
Connolly, K J 142
Consortium, A B N 65
Cords, M 54
Corp, N 50, 61, 81, 102, 103, 143, 145
Crawford, M P 89
Crockford, C 20, 55
Cunningham, E 120
Custance, D M 36

D
Dally, J M 8, 42, 48, 87
Dasser, V 55
Dawkins, R 40
Deacon, T W 22, 164
Deaner, R O 62
de Vries, H 17
de Waal, F B M 54, 57, 85, 86, 98, 99
Dickinson, A 118
Di Fiore, A 120
Douglas-Hamilton, I 96
Dunbar, R I M 53, 54, 59, 60, 106, 109
Dyer, F 122

E
Eddy, T J 41, 46
Elliott, J M 142
Emery, N J 8, 62, 63, 65, 87

F
Feeney, M C 118
Fischer, J 22
Fisher, J 74
Flombaum, J I 87
Fodor, J A 68
Fox, E 143
Fragaszy, D 112, 117, 142
Franks, N R 101

Frith, C D 176
Frith, U 8, 10, 176
Fruth, B 77

G

Galdikas, B M F 113
Galef, B G 71, 79
Gallese, V 97, 139
Gallup, G G 92, 93, 94, 97
Garber, P 120
Gardner, B T 22, 152
Gardner, R A 22, 152
Genty, E 23, 25, 26, 27, 28, 32
Gibson, K R 112
Glickman, S E 131, 132
Goodall, J 77, 80, 82, 143
Goody, E N 61
Gordon, J D 11
Granli, P K 96
Guinet, C 104

H

Hakeem, A Y 107
Hamilton, E 52
Hamilton, W D 58
Hamm, M 32
Happe, F 8
Harcourt, A 53, 54
Hare, B 48, 86
Hauser, M D 100, 101, 102
Healy, S D 65
Heinrich, B 49
Held, S 49
Helfman, G S 79
Helsler, N 22
Hemelrijk, C K 54, 67
Heyes, C M 9, 45, 86, 93, 137, 140
Hill, R A 58
Hinde, R A 74
Hobaiter, C 23, 24, 25, 29, 30, 33, 34, 35, 77
Hohmann, G 77
Hoppitt, W 70, 79, 98
Humle, T 76, 81, 103
Humphrey, N K 56, 57, 109, 130

J

Jabbi, M 97
Jackson, R 120
Janik, V M 137
Janmaat, K R L 119, 120
Janson, C H 120, 121
Janson C H 120
Jarvis, E D 65
Jerison, H J 59, 63
Jolly, A 56
Joly, M 121
Judge, P 55

K

Kamil, A C 117
Kaminski, J 24, 42, 87, 88
Kandel, E R 10
Karin-D'Arcy, M 86
Karmiloff-Smith, A 161
Keysers, C 97, 107
King, B J 25, 112
Kitchen, D M 55
Kobayashi, H 40
Kohshima, S 40
Koops, K 77
Krebs, J R 40
Kuczaj, S 99
Kummer, H 47, 48, 53, 57

L

Laland, K N 62, 70, 71, 75, 79
Lee, P C 112
Lefebvre, L 63, 74
Lehmann, H E 98
Leland, S 100
Lenneberg, E H 163
Liebel, K 23, 24, 26
Limongelli, L 112, 127
Linden, E 22
Loretto, M C 42
Loucks, J 139

M

McAuliffe, K 101
McComb, K 96, 129
McGrew, W C 75, 77, 111, 126
Machiavelli, N 57
McKiggan, H 92
Mackinnon, J 129
MacLeod, C E 158
Macphail, E M 9, 57
Marino, L 93
Marr, D 158
Matsuzawa, T 76, 81, 143
Mead, A P 56
Melis, A P 87, 90
Meltzoff, A N 40
Mercader, J 78
Miles, H L 22, 82, 89
Mitani, J C 82, 89
Mitchell, R W 32
Morgan, D 111
Morris, R G 121
Morton, J 10
Moura, A C de A 142

N

Napier, J R 142, 164
Newell, A 11
Nisbett, R E 69
Noser, R 118, 122, 124

O
O'Neill, D K 89
Onishi, K H 88

P
Parker, S T 112
Parr, L A 98
Patterson, F 22
Paz-y-Miño, G 62
Pepperberg, I M 11
Perry, S 78
Pfungst, O 45
Pika, S 26
Pilbeam, D 94
Pinker, S 163
Platt, M L 41
Plooij, F X 27
Plotnik, J M 90, 93
Polansky, L 121
Poole, J H 86
Povinelli, D J 41, 44, 45, 46, 86, 89, 90, 91, 93, 94
Premack, D 84, 85
Prior, H 93

R
Raihani, N J 101
Rapaport, L G 102, 103, 104
Ray, E D 140
Reader, S M 62, 72
Reid, P J 41
Reiss, D 93
Rendell, L 75
Richardson, T 101
Ridley, A R 101
Riedman, M L 125
Ristau, C 24, 47
Rizzolatti, G 97, 130, 139
Rosch, E 128
Ross, L 69
Rowe, C 65
Ruiz, A M 41, 43, 88
Russon, A E 28, 81, 113, 141, 143

S
Sabbatini, G 112
Saffran, J R 150
Saggerson, A 137
Sambrook, T 53
Santos, L R 87, 88
Sanz, C 111
Savage-Rumbaugh, E S 152
Schel, A M 20
Schultz, E T 79
Seed, A M 62, 98, 127
Seyfarth, R M 15, 16, 17, 18, 54, 55, 61, 62
Sharman, M 58
Shettleworth, S J 13, 53, 118
Shultz, S 60

Shumaker, R 80, 124
Sigg, H 121
Simon, H A 11
Slater, P J B 137
Smet, A F 24
Smith, R 94
Snowdon, C T 103
Southgate, V 88
Sroges, R W 131, 132
Sternberg, R J 61
Stokes, E J 143, 145
Stolba, A 121
Struhsaker, T T 15
Suarez, S A 120
Suzuki, S 111

T
Takasaki, H 72
Tanner, J E 23, 27, 28, 36, 37, 47
Taylor, R J 58
Tebbich, S 125
Tennie, C 148
Terborgh, J 59
Thompson, E 97
Thorndike, E L 136
Thornton, A 101
Tomasello, M 23, 26, 29, 41, 47, 48, 85, 86, 87, 90, 117, 145, 148, 155
Tulving, E 118
Tutin C E G 77

U
Udell, M A R 42

V
van Roosmalen, A 54, 98
van Schaik, C P 58, 70, 75, 79, 80
Vasey, P 113
Venditti, C 158
Viranyi, Z 42
Visalberghi, E 112
Vonk, J 86
Voronov, L N 65

W
Wallman, J 14
Warneken, F 90
Warner, R R 79
Warren, J M 57, 59
Washburn, D A 55
Watts, D P 82, 89
Watve, M 42, 47
Whitehead, H 75
Whiten, A 26, 47, 50, 53, 55, 56, 57, 60, 61, 66, 67, 75, 76, 86, 88, 89, 141, 149
Wich, S A 17
Wicker, B 97, 120
Wilkinson, A 42

Willems, E P 58
Woodruff, G 84, 85
Wrangham, R 64

Y
Yamakoshi, G 143
Young, R M 11

Z
Zacks, J M 139
Zentall, T R 137
Zimmermann, E 121
Zuberbühler, K
   16, 19, 27

Subject index

A
actions:
　elaborate, complex scheduling 158
　imitation at level of 140–1
　repertoire *see* repertoire of actions
　segmenting streams of 138–40
　see also motor acts; skills
aggregations of different species 59
aggression, primate 55
alarm calls/vocalizations 20, 58
　primates 15–17, 20, 62, 127–8
　starlings 18
Alex, Pepperberg's gray parrot (*Psittacus erithacus*) 11
alliances and coalitions 53, 54, 62, 65, 66, 67
allometry 4, 63, 106
American Sign Language 20, 22
ant(s) 115
　chimpanzees eating 75–6, 115
　teaching each other 101
anthropology 76, 82, 127, 164
apes *see* bonobo, chimpanzee, gorilla, orangutan
apparently satisfactory outcome (ASO), in gestural communication studies 33
associative learning 5, 6, 9, 113
audible communication *see* vocal communication
audience targeting in gestural communication 23–5

B
babbler, pied (*Turdoides bicolor*)
　teaching 101, 102
baboons (*Papio* sp.):
　aggregations with impala 59
　cognitive maps 121, 124
　feeding 118–19
　social complexity 53, 55–6
　understanding gaze 53
　understanding others 47, 48
bee-eater, little (*Merops orientalis*) 42, 44, 50
behavior (animal):
　cognition and 9–13
　parsing *see* parsing
birds:
　brain size 64–5
　gaze-following 42
　memory of place 117–18
　mirror self-recognition 91
　social learning 74–5
bird tables 74
body size 106, 114
　and brain size 4, 59, 63, 64, 106, 157
bonobo (*Pan paniscus*):
　cultural intelligence 77
　gestural communication 26, 27, 30
　parsing human speech 152
　social abilities 109, 110
　technical abilities 110
Bossou (Guinea), chimpanzees 76
brachiation 114
brain 61–5
　cognitive demands and changes in 105
　mentalizing and the 107
　priming *see* priming
　size 4, 59–60, 61, 63–5, 108, 156–7, 163
　social 59–63
　see also cerebellum, neocortex
Budongo Forest, chimpanzee community 20, 29–30

C
captivity, *see* zoo animals
capuchin monkeys 78, 112, 121, 127, 140
cats:
　mirror self-recognition 91
　teaching 101
causes, understanding/"seeing" 165–6
　parsing and 159–60
　tool use and 124–7
Cercopithecine monkeys 114
cerebellum 59, 158, 164, 167
chaffinch (*Fringilla coelebs*) 31
cheetah (*Acinonyx jubatus*), teaching by 101, 101–2
chickadees (tits) 117, 118, 119
chimpanzee (*Pan troglodytes*) 1–2
　culture and skills learning 80–1, 82
　fruit eating 72, 81, 102–3
　gestural communication 20, 22–38
　learning from others 75–82
　numerical abilities 11
　nut cracking 78, 81, 102, 103, 112, 115, 136
　Sultan (Köhler's) 1, 2
　tool use 125–6, 132
　　termite fishing 76, 80, 82, 110–11, 115, 126

SUBJECT INDEX

chimpanzee (*continued*)
  understanding what others see, know, and think 44–49, 87–8, 89–91, 92, 93, 97, 98–9, 102, 103, 105
  vocal communication 19, 20
Chlorocebus pygerythrus see vervet monkey
chunking of actions 138–40
Clark's nutcracker (*Nucifraga columbiana*) 117
Clever Hans 45–6
coalitions and alliances 53, 54, 62, 65, 66, 67
cognition 68
  advanced/sophisticated 13, 62, 65, 76, 155
  behavior and 9–13
  brain changes and cognitive need 105
  without much insight 155–7
cognitive maps 120–4
collaboration, understanding another's role 89–91
communication:
  gestural *see* gestural communication
  social 39
  vocal *see* language; speech; vocal communication
competition 8, 48–9, 50, 54, 68, 86, 87, 118, 156
  groups and 58, 59
complex skills learning 7, 80–1, 134–54
consciousness 12–13
conservatism (in social learning) 71, 72, 73, 83, 156
contextual imitation 137, 141
cooperation, understanding another's role 89–91
copying *see* imitation and copying
Corvus frugilegus see rook
culture 75–83
curiosity 130–2

D

danger and risk categorization 127–9
death, understanding 93, 95–6
deception, *incl.* tactical deception 46, 50, 55–6, 60, 61, 67, 72–3, 88, 156
deer, red (*Cervus elaphus*) 95
diet 77
  brain size and 64
  see also food and feeding
digit role differentiation 142–3
dog (*Canis familiaris*), gaze-following 41–2
dolphins:
  empathy 99
  mirror self-recognition 93

E

eating *see* diet; food and feeding
educating *see* teaching, learning
elephants:
  empathy 99

gestural communication 24
mirror self-recognition 93
risk and danger categorization 128–9
teaching 103–4
understanding death 95–6
emotional contagion 98
empathy 97–100
Euclidean properties of cognitive maps 120, 122–3, 124
exclusion method for identifying animal culture 26, 75, 76–8
exploration, individual 70, 71
  trial-and-error learning and 5, 67, 79, 136, 137, 141
extinct apes 114–15
eyes:
  gaze *see* gaze
  sclera 40, 46
  see also seeing; visual gestures

F

facial expressions 23
false beliefs 86, 87, 88–89, 105, 160
feeding *see* food and feeding
food and feeding/eating 7, 72, 110–16
  foraging 43, 44, 112, 114, 117
  gaze and 43, 44
  learning from others 74–5, 79, 80–1, 83
  memory for food-storage 117–18
  tool use 78, 80, 81, 102, 103, 110–13, 115, 116, 125, 126, 127, 136
  see also diet, primates
friendships 53, 54, 156
fruit-eating primates 72, 80, 81, 102–3, 113, 114, 116, 118, 119
functional reference 17–18, 19, 20

G

g (general intelligence) 61, 62, 63
Galapagos woodpecker finch (*Camarhynchus pallidus*) 80, 125
gaze 7, 41–4
  following 40, 41–2, 43, 46, 49–50, 68
  priming 44, 50, 68
general intelligence (g) 61, 62, 63
gestural communication (in apes) 7, 20, 23–39, 163
  audience targeting 23–5
  imitation in 35–7
  intended meanings 31–4
  in play 25–6
  repertoire development 26–31
  strings of gestures 34–5
gibbons 94, 114
goats (*Capra hircus*), gaze-following 42
Gombe (Tanzania), chimpanzees 75–6, 76, 102

gorilla *incl.* mountain gorilla (*Gorilla beringei beringei*) and western gorilla (*Gorilla gorilla*):
  deception 55
  gestural communication 23, 24, 26, 27, 28, 29–30, 31, 32, 33, 36, 36–7
  imitation 142–54
  technical abilities 110, 111, 112–13
  reaction to death 95–6
  understanding others 47
Goualougo chimpanzees 111, 126
great apes *see* primates
grooming, social 53–4
groups 58–9, 60, 109
  brain and 60, 63
  identity 82
guesser–knower experiment 44–5, 46, 49, 86

H
hamadryas baboon *see* baboons
hammer and anvil technique 78, 81, 82, 103, 112, 115, 125, 127
hippopotamus, empathy 100
honey:
  chimpanzees and 111
  orangutans and 80, 116
horse, understanding others 45–6
humans:
  culture, uniqueness 83
  risk and danger categorization by 128
  risk and danger categorization of, by elephants 128–9

I
iconic gestures 27, 28, 31
ideas 17, 165–6
ignorance 75
  understanding 49, 50, 51, 105
imitation and copying 134–54
  action-level 140–1
  contextual 137, 141
  defining and recognizing 136, 149
  gestural 35–7
  managing without 135–6
  production 137–8, 142
  program-level 141–2, 148
impala (*Aepyceros melampus*):
  aggregations with baboons 59
  empathy 100
innovation and invention (new skills) 71, 72–3, 74, 75, 76, 78, 83
insight, defining 1–6
intelligence:
  cultural 75–83
  general (g) 61, 62, 63
  social 53, 56–7, 61–3, 65–8, 109, 110, 110
intentionality 161, 164, 165, 166, 167
  in gestural communication 23, 27, 29, 31
  parsing and 159–60

teaching 102, 104
theory of mind and 85, 102
in vocal communication 17–18
intricate complexity in skills 7, 80–1, 138
invention and innovation (new skills) 71, 72–3, 74, 75, 76, 78, 83
isocortex *see* neocortex

J
jay, *see* scrub jay, western
joint activity, understanding another's role 89–91

K
Kanzi (bonobo) 152
Karisoke (Rwanda) mountain gorillas 142, 144, 146, 148
knowledge, *see also* social knowledge; understanding
Köhler, chimpanzee Sultan 1, 2

L
language 3, 7, 14
  ape 38, 152, 163, 164
  evolution 163–4
  sign 20, 22
  see also speech
langur, Thomas's (*Presbytis thomasi*) 17
learning:
  associative 5, 6, 9, 113
  gestures 26–7
  by imitation *see* imitation
  from others 70–83
  rapid 61, 66–7, 86, 105, 162
  of skills *see* skills
  social 70–2, 73, 74, 75, 79, 80, 83, 125–6, 134, 135, 136, 137, 156
  see also teaching
lemurs:
  gaze 41, 43, 44
  social intelligence 56
leopard (*Panthera pardus*) 16, 19, 95, 127–8
locations (places), memory of 117–19
long-term memory 67, 117–19

M
Maasai and elephant risk categorization 128–9
macaque monkeys (*Macaca* sp.):
  curiosity 130–1
  empathy 97, 98
Machiavellian intelligence 57
magpie (*Pica pica*), mirror self-recognition 93
mangabey monkeys (*Cercocebus* sp.) 119, 120
mango eating, Mahale chimpanzees 72
manual role differentiation 142, 143
maps, cognitive 120–4
meanings, intended 31–4
meerkat (*Suricata suricatta*), teaching 101, 102

memory 67, 117–19
  long-term 67, 117–19
mentalizing 85, 106, 160, 166
mental representation 2–4, 5, 6, 157–8
  physical world and 119, 120, 123, 124, 132, 133
mental state, insight into *see* theory of mind
Merops orientalis see bee-eater, little
miming actions 28
Miocene apes 115
mirror, self-recognition 91–4
mirror neurons 97, 140
monkey *see* primates *and individual species*
monkey see, monkey do cells (mirror neurons) 97, 140
mortality (death), understanding 93, 95–6
motor acts and behavior parsing 149

N
navigation 120–4
Ndoki chimpanzees, termite eating 110–11
Neesia fruit 80
neocortex (isocortex) 50, 56, 59, 60, 61, 62, 68, 72, 109, 157, 158, 167
  bird brain and 65
nettle-eating, by gorillas 142–52, 154
network-map 123–4
neurons 65
  interconnectedness 63
  mirror 97, 140
  spindle (von Economo) 107
non-verbal tests of theory of mind 161
numerical abilities of animals 11
nutcracker, Clark's (*Nucifraga columbiana*) 117
nut-cracking primates 78, 81, 102, 103, 112, 115, 136

O
Ockham's razor 52
olfaction 64, 103, 129
ontogeny:
  gesture 37
  by ritualization 27–31, 38
orangutan (*Pongo* sp.):
  culture 75, 76–7, 78, 79, 80, 81
  gestural communication 24–5, 26, 28, 29, 30, 31, 32, 33, 36
  social abilities (incl. learning from others) 75, 76–7, 78, 79, 80, 81, 109–10
  technical abilities 109, 111, 113, 116, 126–7, 143
  understanding others' thoughts 92, 93–4
orca (*Orcinus orca*) 104
others:
  learning from 70–83
  understanding *see* understanding

P
Pan see bonobo, chimpanzee
Pandora (mountain gorilla) 146
Panthera pardus see leopard
Papio papio see baboons
parrot abilities 11, 63
parsimony 10, 52, 53
parsing:
  behavior 7, 148–54, 158–63, 164, 166
    extraction of statistical regularities 149–52, 166
    to "see" causes and intention 159–60
  speech, bonobo 152
Pepperberg, gray parrot Alex 11
perception, sophisticated 66, 67, 68, 156
Pfungst, horse Clever Hans 45–6
physical world, knowledge of 7, 117–33
  primates 118–19, 120–1, 121, 124, 157
Picasso (mountain gorilla) 148
pied babbler (*Turdoides bicolor*), teaching 101, 102
pigs 49, 51
piping plover (*Charadrius melodus*) 24, 31, 47
places, memory of 117–19
play:
  gesturing in 25–6
  understanding others 47
plover, piping (*Charadrius melodus*) 24, 31, 47
predator avoidance 62, 127
  alarm calls *see* alarm calls
  groups and 58
primates (non-human incl. monkeys and great apes) 157–67
  behavior parsing 7, 148–54, 158–63, 164, 166
  cognition without much insight 156–7
  culture 75–83
  curiosity 130–1
  extinct 114–15
  feeding 110–16
    fruit 72, 80, 81, 102–3, 113, 114, 116, 118, 119
    nettles, eating by gorillas 142–52, 154
    nut-cracking 78, 81, 102, 103, 112, 115, 136
  gestural communication *see* gestural communication
  imitation 138, 142–54
  physical world knowledge 118–19, 120–1, 121, 124, 157
  primates 118–19, 120–1, 121, 124, 157
  skills learning 79, 80–2, 142–54
  social behavior 52–69
  tactical deception 46, 50, 56, 60, 67, 72–3, 88
  technical abilities/tool use 78, 80, 109–16, 126–7, 132
  understanding thoughts of others 86–107

vocal communication *see* language; speech; vocal communication
see also bonobo; chimpanzee; gorilla; orangutan
priming 16, 71, 136
  gaze 44, 50, 68
production imitation 137–8, 142
program-level imitation 141–2, 148

R

rat (*Rattus norwegicus*):
  cognitive maps 121, 122, 124
  culture 79
  curiosity 130
  laboratory study of learning 9
  social learning 71
raven (*Corvus corax*) 48, 49, 50, 51, 64, 68, 87, 105, 106
red deer (*Cervus elaphus*) 95
reference 16
  functional 17–18, 19, 20
  idea of 165
repertoire of actions 134, 135, 137, 138–40, 143
  gestures 26–31, 38
  infant gorilla 149
  observing 139
representation *see* mental representation
response facilitation 71, 135–6, 137, 140
rhesus monkey (*Macaca mulatta*):
  mirror neurons 139–40
  understanding others' thoughts 87, 88
risk and danger categorization 127–9
ritualisation 29
  ontogenetic 27–8
rook (*Corvus frugilegus*) 62, 98, 105, 106
route maps 123, 124
Rwanda (Karisoke) mountain gorillas 142, 144, 146, 148

S

Saba fruit 81, 102
sclera 40, 46
scrub jay, western (*Aphelocoma californica*) 8, 42, 48, 50, 62, 87, 105, 106, 117, 118, 119
seeing, knowing about 44–9, 71
  see also eyes; visual gestures, silent
segmenting streams of actions 138–40
self-recognition, mirror 91–4
Sign Language, American 20, 22
silent visual gestures 23
skills (and their learning) 80–2, 134–54
  complex new 7, 80–1, 134–54
  innovation and invention of new skills 71, 72–3, 74, 75, 76, 78, 83
  primate 79, 80–2, 142–54
  social *see* social skills
  see also actions, tool use
smell sensation (olfaction) 64, 103, 129

social behavior (in general) 52–69
social communication 39
social complexity 53–4
social insight 7, 108
social intelligence 53, 56–7, 61–3, 65–8, 109, 110, 110
social knowledge 54–6, 68, 84
social learning 70–2, 73, 74, 75, 79, 80, 83, 125–6, 134, 135, 136, 137, 156
social skills/abilities:
  brain and 59–63
  vs. technical abilities 108–17
Sonso (Budongo Forest) chimpanzee community 20, 29–30
space and navigation 120–4
speech (human) 163
  bonobo parsing 152
  see also language
spindle neurons 107
starling, superb (*Lamprotornis superbus*) 16, 18
statistical regularities, extraction of 149–52, 166
stimulus enhancement 70–1, 135–6, 137
stinging nettle, eating by gorillas 142–52, 154
Sultan (Köhler's chimpanzee) 1, 2
sympathetic concern 98, 99, 105, 106
synonyms, gestural 35

T

tactical deception 46, 50, 55, 56, 60, 67, 72–3, 88
tactile gestures 23–4
Taï (Ivory Coast) chimpanzees 76, 78, 102, 103
tamarin monkeys 103
teaching 102–4
  see also learning
technical abilities *see* tool use
termites 77
  fishing 76, 80, 82, 110–11, 115, 126
theory of mind (and understanding others) 7, 8, 40, 84–107
  non-verbal tests of theory of mind 161
thinking 12
  undertanding others *see* theory of mind
Thomas's langur (*Presbytis thomasi*) 17
tits (chickadees) 117, 118, 119
tool use (technical abilities) 78, 80, 124–7
  feeding 78, 80, 81, 102, 103, 110–13, 115, 116, 125, 126–7, 136
  primates 78, 80, 109–16, 126–7, 132
traditions, animal 73–5, 137
trial-and-error learning and exploration 5, 67, 79, 136, 137, 141

U

unconscious processes 166
understanding and knowledge:
  of oneself 91–4

understanding and knowledge (*continued*)
  of others (knowing about them) 39, 69
    their seeing 44–9, 71
    their thinking *see* theory of mind
  of physical world *see* physical world
ungulates 46, 59, 60

V

vervet monkey (*Chlorocebus pygerythrus*):
  danger and risk reactions 127–8
  social intelligence 61–2
  vocal communication 15, 16, 18, 19
visual gestures, silent 23
  see also eyes; seeing
vocal (audible) communication, primates 7, 14–21
  alarm 15–17, 20, 62, 127–8
  gestures 24
  see also language; speech
voluntary control of ape gestures 23
von Economo neurons 107

W

whales, empathy 99–100
wolf (*Canis lupus*), gaze-following 41–2
woodpecker finch, Galapagos (*Camarhynchus pallidus*) 80, 125

Z

zoo animals:
  curiosity 131, 132
  gestural communication 25, 26, 27, 28, 32, 33, 36
  see also captivity
Zura (gorilla) 36–7